A Carasso & A P Stone (Editors)
University of New Mexico

Improperly posed boundary value problems

This collection of papers
is dedicated to Fritz John

Pitman Publishing
LONDON · SAN FRANCISCO · MELBOURNE

AMS Subject Classifications: 35R25, 35R30, 65M30

PITMAN PUBLISHING
Pitman House, 39 Parker Street, London WC2B 5PB, UK

PITMAN PUBLISHING CORPORATION
6 Davis Drive, Belmont, California 94002, USA

PITMAN PUBLISHING PTY LTD
Pitman House, 158 Bouverie Street, Carlton, Victoria 3053, Australia

PITMAN PUBLISHING
COPP CLARK PUBLISHING
517 Wellington Street West, Toronto M5V 1G1, Canada

SIR ISAAC PITMAN AND SONS LTD
Banda Street, PO Box 46038, Nairobi, Kenya

PITMAN PUBLISHING CO SA (PTY) LTD
Craighall Mews, Jan Smuts Avenue, Craighall Park,
Johannesburg 2001, South Africa

ISBN 273 001051

Reproduced and printed by photolithography and bound in Great Britain at The Pitman Press, Bath

3504 : 97

Preface

In May 1974, a five-day regional conference on the subject of "Improperly Posed Problems in Partial Differential Equations" was held at the University of New Mexico, Albuquerque, NM, with the support of the National Science Foundation. The conference attracted participants from a broad geographical region within the U.S. The principal speaker was L.E. Payne of Cornell University. The lectures by Professor Payne entitled "Improperly Posed Problems in Partial Differential Equations" are to be published by SIAM in its CBMS Regional Conference Series in Applied Mathematics.

In recent years, the general topic of ill-posed problems has interested an increasing number of mathematicians and a fair body of knowledge has been accumulated. Moreover, as applied mathematicians tackle the increasingly complex problems posed by modern advances into such fields as meterology, geophysics, fluid dynamics, elasticity, the life sciences, and other areas, the likelihood of encountering such problems increases. Consequently, a conference devoted to this special chapter in the general theory of partial differential equations seemed appropriate.

In organising this conference, the Editors kept in mind that several of the invited participants had themselves made original contributions to the field. Accordingly, it was possible to provide a fairly broad coverage of the area, by supplementing the morning lectures by the principal speaker with afternoon lectures by the participants. The present collection of papers comprises almost all of the afternoon lectures. The Editors take this opportunity to thank the participants and Professor Payne for their efforts in making this conference such a stimulating event.

To our knowledge, this conference was the first to be

held in the United States on the subject of ill-posed problems. On this occasion, it is especially fitting to dedicate this collection of papers to Fritz John, in recognition of his pioneering and fundamental contributions to this area of analysis over the last quarter of a century.

Albuquerque, New Mexico
August 1975

Alfred Carasso
Alex P. Stone

Contents

H F WALKER

Well-Posedness of Certain Elliptical Problems in Unbounded Domains

1. INTRODUCTION

In this note, we discuss the existence and uniqueness of solutions $u \in H_{m,p}(R^n; C^k)$ of an mth- order k-dimensional system of partial differential equations

$$Au = f, \ f \in L_p(R^n; C^k), \tag{1}$$

where the operator $A = \sum_{|\alpha| \leq m} a_\alpha(x) \frac{\partial^\alpha}{\partial x^\alpha}$ is <u>elliptic</u>, ie.,

$\sum_{|\alpha|=m} a_\alpha(x)\xi^\alpha$ is a non-singular k X k matrix for all $x \in R^n$ and all non-zero $\xi \in R^n$. $L_p(R^n; C^k)$ and $H_{m,p}(R^n; C^k)$ are the usual Banach spaces of C^k-valued functions on R^n, denoted henceforth by L_p and $H_{m,p}$, respectively. The notation is standard multi-index notation. In the following, we always assume that $1 < p < \infty$ and $1/p + 1/q = 1$.

In classical investigations, in which the independent variables are restricted to a compact set or manifold and appropriate boundary conditions are imposed, it is shown that an elliptic operator A is <u>Fredholm</u> on L_p, i.e., R(A) (the range of A in L_p) is closed and both the dimension of N(A) (the null-space of A in L_p) and the co-dimension of R(A) (the dimension of N(A*) in L_q) are finite. (For questions concerning the definition and basic properties of Fredholm operators, see [1].). Thus questions of existence and uniqueness can be answered in this case by examining the finite - dimensional subspaces $N(A) \subseteq L_p$ and $N(A^*) \subseteq L_q$. Furthermore, the property of being Fredholm is preserved under small perturbations of the operator, and the <u>index</u>, defined by Ind A = dim N(A) - dim N(A*), is a homotopy invariant of the operator.

In the case at hand, the independent variables range over all of R^n, and an elliptic operator cannot be Fredholm on L_p. (In fact, the range of an elliptic operator with constant coefficients is dense but not closed in L_p.) However, it is seen in our discussion of existence and uniqueness of solutions of (1) that, subject to the restrictions described below, such an operator is "practically" Fredholm in the following sense: Not only does such an operator have a finite-dimensional null-space, but it also has a finite, perturbation-invariant index and its range can be characterized in terms of the (known) range of an operator with constant coefficients and a finite, index-related number of orthogonality conditions.

The "uniqueness" discussion below is entirely a summary of material which has already appeared in print ([3], [4], [6].) Consequently, no proofs are given for the "uniqueness" results. The "existence" discussion is modelled along the lines of [5], but is primarily a report on work in progress.

2. UNIQUENESS

Questions about the uniqueness of solutions of (1) can be answered by an examination of N(A), the null-space of A in L_p. Such null-spaces are rather exhaustively studied in [3], [4], and [6] for an operator A which can be written in the form $A = A_\infty + B$, where $A_\infty = \sum_{|\alpha|=m} a_\alpha \frac{\partial^\alpha}{\partial x^\alpha}$ is an elliptic $m^{\underline{th}}$ - order homogeneous operator with constant coefficients and $B = \sum_{|a|\le m} b_\alpha(x)\frac{\partial^\alpha}{\partial x^\alpha}$ is an $m^{\underline{th}}$ - order perturbation term. The following is the main theorem of [3].

<u>Theorem 1</u>: Suppose that, for some $\delta > 0$,

$$\lim_{|x|\to\infty} \sup \; |x|^{m-|\alpha|} \; |b_\alpha(x)| < \delta$$

for all $|\alpha| \le m$. If δ is sufficiently small (the exact smallness depending on the coefficients of A near the origin in R^n) then dim $N(A) < \infty$.

The theory of Fredholm and semi-Fredholm operators leads one to hope that dim N(A) might behave upper-semicontinuously under perturbations of the coefficients of A. In fact, under no more assumptions on A than in the above theorem, it is shown in [3] that dim N(A) depends upper-semicontinuously on A if and only if $m - \frac{n}{q}$ is not a non-negative integer. If $m - \frac{n}{q}$ is a non-negative integer, then dim N(A) behaves upper-semicontinuously only if stronger assumptions are made about the coefficients of the perturbation term B, e.g. $\sup_{\substack{x \in R^n \\ |\alpha| \leq m}} |x|^{m-|\alpha|+\rho} |b_\alpha(x)| < \infty$ for $\rho > 0$. This dependence on $m - \frac{n}{q}$ arises from the crucial inequality

$$\sum_{|\alpha| \leq m} \left\| |x|^{|\alpha|+\rho} \frac{\partial^\alpha}{\partial x^\alpha} u \right\|_p \leq C \, \| |x|^{m+\rho} A_\infty u \|_p, \qquad (2)$$

a necessary condition for which is that $\rho + m - \frac{n}{q}$ not equal a non-negative integer.

Another inequality used in establishing finite-dimensionality of N(A) and upper-semicontinuity of dim N(A) is the following: If there exists a (sufficiently small) $\delta > 0$ such that

$\limsup_{|x| \to \infty} |b_\alpha(x)| < \delta$ for $|\alpha| = m$ and if

$\sup_{x \in R^n} |x|^{m-|\alpha|} |b_\alpha(x)| < \infty$ for $|\alpha| < m$, then, for all real ρ,

$$\sum_{|\alpha| \leq m} \left\| |x|^{|\alpha|+\rho} \frac{\partial^\alpha}{\partial x^\alpha} u \right\|_p \leq C\{ \| |x|^{m+\rho} A u \|_p + \| |x|^\rho u \|_p \}. \quad (3)$$

Inequalities (2) and (3) play the same role here as that of the usual coerciveness inequalities in other investigations.

In obtaining the inequality (2), use is made of the following lemma about integral operators. This lemma is used in the "existence" discussion below.

<u>Lemma 1</u>: For real a and b with $a + b > 0$, the integral operator

$$Ku(x) = \int_{R^n} \frac{1}{|x|^a \, |x-y|^{n-a-b} \, |y|^b} u(y) \, dy$$

is bounded on L_p if and only if $a < \frac{n}{p}$ and $b < \frac{n}{q}$.

3. EXISTENCE

To answer questions about the existence of solutions of (1), we show that, at least in the case considered, the operator $A = A_\infty + B$ has a finite, perturbation-invariant index and the range of A can be characterized in terms of the (known) range of A_∞ and a finite, index-related number of orthogonality conditions. We will develop only the L_2-theory of first-order operators when the number of independent variables is at least three ($p = 2$, $m = 1$, $n \geq 3$). There are two reasons for not dealing in greater generality: First, the details of the general case have not been worked out yet, although doing so appears to be straightforward; second, the additional complexity of the general case obscures the similarity between our operators and operators which are Fredholm in the classical sense.

We write A as

$$A = A_\infty + B = \sum_{i=1}^{n} a_i \frac{\partial}{\partial x_i} + \sum_{i=1}^{n} b_i(x) \frac{\partial}{\partial x_i} + b_0(x).$$

Since the estimate (2) holds with $\rho = 0$, we assume

$$\lim_{|x| \to \infty} |b_i(x)| = 0, \; i = 1, \ldots, n, \text{ and } \lim_{|x| \to \infty} |x| \; |b_0(x)| = 0.$$

(Actually, we could carry out everything that follows assuming only that $\limsup_{|x| \to \infty} |b_i(x)|$, $i = 1, \ldots, n$, and $\limsup_{|x| \to \infty} |x| \; |b_0(x)|$ are small. However, it is more convenient to assume that these quantities vanish.). Finally, we assume that the functions $b_i(x)$ are differentiable and satisfy $\lim_{|x| \to \infty} |x| \; |\frac{\partial}{\partial x_i} b_i(x)| = 0.$

This is not needed for all of the results below, but it is used in obtaining the final result. In fact, since

$$A^* = A^*_\infty - \sum_{i=1}^{n} b_i^* \frac{\partial}{\partial x_i} - \sum_{i=1}^{n} \frac{\partial}{\partial x_i} b_i^* + b_0^*,$$

one sees that this assumption merely guarantees that A^* has the same properties as A.

One would suspect that the greatest effects of the perturbation term B occur near the origin. Our plan is to isolate these effects in a certain sense. Toward this end, let $\phi_1 \in C^\infty(R^n)$ satisfy $\phi_1(x) = 0$ for $|x| \leq 1/2$ and $\phi_1(x) = 1$ for $|x| \geq 1$. For $r > 0$, set $\phi_r(x) = \phi_1(x/r)$ and define $A_r = A_\infty + \phi_r B$. Set

$$M_r = \{u \in H_{1,2} : Au(x) = 0 \text{ if } |x| \geq r\},$$

and let D_r denote the ball of radius r about the origin in R^n. Our fundamental result is the following theorem.

Theorem 2: For large r, $A: M_r \to L_2(D_r; C^k)$ is a bounded Fredholm operator. Its null-space is $N(A)$, and the orthogonal complement of its range in $L_2(D_r; C^k)$ is the restriction of functions in $N(A^*)$ to D_r.

It follows from this theorem and the basic properties of Fredholm operators that dim $N(A)$ - dim $N(A^*)$ is finite and invariant under small perturbations of the coefficients of A inside D_r. Thus we define the index of A by Ind A = dim $N(A)$ - dim $N(A^*)$. Before proving the theorem, we establish two lemmas which enable us to obtain the desired characterization of $R(A)$ as a corollary to the theorem.

Lemma 2. $B u \in R(A_\infty)$ for all $u \in H_{1,2}$.

Proof. We will show the existence of an inequality $||B^*\phi||_2 \leq C \, ||A_\infty{}^*\phi||_2$ for all $\phi \in C_0^\infty(R^n; C^k)$. From this, one sees that, for each $u \in H_{1,2}$, there exists a $v \in L_2$ such that $(Bu, \phi) = (v, A_\infty^*\phi)$ for all $\phi \in C_0^\infty(R^n; C^k)$. It follows that $v \in H_{1,2}$ and $A_\infty v = Bu$.

We have that $B^* = -\sum_{i=1}^n b_i^* \frac{\partial}{\partial x_i} - \sum_{i=1}^n \frac{\partial}{\partial x_i} b_i^* + b_0^*$. First, the ellipticity of A_∞^* implies that

$$||b_i^* \frac{\partial}{\partial x_i} \phi ||_2 \leq C \; || \frac{\partial}{\partial x_i} \phi ||_2 \leq C \; ||A_\infty^*\phi||_2, \quad i = 1, \ldots, n$$

for $\phi \in C_0^\infty(R^n; C^k)$. Second, letting Γ be the fundamental solution of A_∞, we have

$$|(- \sum_{i=1}^n \frac{\partial}{\partial x_i} b_i^*(x) + b_0^*(x))\phi(x)| \leq \frac{C}{|x|} \; | \int_{R^n} \Gamma^*(y-x) A_\infty^*\phi(y)\,dy|$$
$$\leq C \int_{R^n} \frac{1}{|x| \; |x-y|^{n-1}} \; |A_\infty^*\phi(y)| \; dy$$

for $\phi \in C_0^\infty(R^n; C^k)$. (For questions concerning the existence and properties of fundamental solutions, see [2].) Then Lemma 1 implies that

$$|| - \sum_{i=1}^n (\frac{\partial}{\partial x_i} b_i^* + b_0^*)\phi||_2 \leq C \; ||A_\infty^*\phi||_2$$

for $\phi \in C_o^\infty (R^n; C^k)$, and the lemma follows.

Lemma 3: For large r, there exist positive constants C_1 and C_2 such that

$$C_1 ||A_r^* u||_2 \leq ||A_\infty^* u||_2 \leq C_2 \ ||A_r^* u||_2$$

for all $u \in H_{1,2}$.

Proof. Note that there exists a positive constant C, independent of r, such that $|\frac{\partial}{\partial x_i} \phi_r(x)| \leq \frac{C}{|x|}$, i = 1, ... for all non-zero $x \in R^n$. Then use Lemma 1 as in the proof of Lemma 2 to derive an inequality

$$||(A_\infty^* - A_r^*)\phi||_2 \leq \epsilon(r) \ ||A_\infty^*\phi||_2$$

for $\phi \in C_o^\infty (R^n; C^k)$, where $\epsilon(r)$ approaches zero as r grows large. The lemma follows easily from this inequality.

Corollary: For large r, $R(A_r) = R(A_\infty)$.

Proof: Suppose $f \in R(A_r)$, i.e., $A_r u = f$ for some $u \in H_{1,2}$. Then for all $\phi \in C_o^\infty (R^n; C^k)$, we have

$$|(f,\phi)| = |(u, A_r^*\phi)| \leq C||A_r^*\phi||_2 \leq C \ ||A_\infty^*\phi||_2$$

by Lemma 3. The Riesz Representation Theorem implies that there exists a $v \in L_2$ such that $(f,\phi) = (v, A_\infty^*\phi)$ for all $\phi \in C_o^\infty(R^n; C^k)$. It follows that $v \in H_{1,2}$ and $A_\infty v = f$. Thus $R(A_r) \subseteq R(A_\infty)$. Similarly, $R(A_\infty) \subseteq R(A_r)$.

The following result is our characterization of R(A).

Corollary to Theorem 2: A function $f \in L_2$ is in R(A) if and only if $f \in R(A_\infty)$ and $f \in N(A^*)^\perp$.

Proof: Let χ_r be the characteristic function of D_r, ie.,

$$\chi_r(x) = \begin{cases} 1 \text{ if } |x| \leq r \\ 0 \text{ if } |x| > r \end{cases}$$

It follows from the corollary to Lemma 3 that, if r is large, then the following are equivalent:

(i) $f \in R(A_\infty)$ and $f \in N(A^*)^\perp$,

(ii) $(1-\chi_r)f \in R(A_r)$ and $[f - AA_r^{-1}(1-\chi_r)f] \in N(A^*)^{\perp}$.

A straightforward application of Lemma 2, Lemma 3, and Theorem 2 shows that (ii) holds if and only if $f \in R(A)$.

Proof of Theorem 2: Clearly, the null-space of $A: M_r \to L_2(D_r; C^k)$ is $N(A)$ for all r. To complete the proof, we will show that, if r is large, then each $v \in L_2(D_r; C^k) \cap A(M_r)^{\perp}$ can be uniquely extended to an element of $N(A^*)$.

Define $L_r = \{u \in H_{1,2}: A_r u = 0 \text{ for } |x| \leq r\}$. If r is large, then each $u \in H_{1,2}$ can be uniquely written as $u = u_1 + u_2$, where $u_1 \in L_r$ and $u_2 \in M_r$. (Indeed, for large r, $\chi_r A_r u \in R(A_\infty) = R(A_r)$ and, therefore, $(1-\chi_r) A_r u \in R(A_r)$. so we take $u_1 = A_r^{-1}(1-\chi_r)A_r u$ and $u_2 = A_r^{-1}\chi_r A_r u$). We also observe that, if r is large, then the following inequalities hold for $u \in H_{1,2}$:

(i) $\|A_\infty u\| \leq C \|A_r u\|$,

(ii) $\|Bu\| \leq C \|A_\infty u\|$ and, hence, $\|Au\| \leq C \|A_\infty u\|$.

The proofs of inequalities (i) and (ii) parallel those of Lemma 3 and Lemma 2, respectively.

Now suppose that $v \in L_2(D_r;C^k) \cap A(M_r)^{\perp}$ and that r is large. For all $u \in H_{1,2}$, we have

$$|(Au,v)| \leq \|v\| \, \|Au\| \leq C \|A_\infty u\| \leq C \|A_r u\|$$

by (i) and (ii) above. It follows, in particular, that the assignment $A_r u \to -(Au,v)$ is a bounded linear functional on $A_r(L_r)$, which, by Lemma 3, is a dense subset of $L_2(R^n-D_r;C^k)$. Consequently, there exists a unique $w \in L_2(R^n-D_r; C^k)$ such that $(A_r u,w) = -(Au,v)$ for all $u \in L_r$.

Let u be an arbitrary element of $H_{1,2}$. Write $u=u_1 + u_2$, where $u_1 \in L_r$ and $u_2 \in M_r$. We have

$$(Au,v+w) = (Au_1,v) + (Au_1,w) + (Au_2,v) + (Au_2,w).$$

The first and second terms cancel, since $u_1 \in L_r$ and $(Au_1, w) = (A_r u_1, w)$. The third term is zero, since $u_2 \in M_r$ and $v \in A(M_r)^{\perp}$. The fourth term is zero, since $u_2 \in M_r$ and $w \in L_2(R^n - D_r; C^k)$. Thus $(Au, v+w) = 0$ for all $u \in H_{1,2}$, and it follows that $v + w$ is the unique extension of v to an element of $N(A^*)$.

REFERENCES

1. I.C. Gohberg and M.G. Krein, The basic propositions on defect numbers, root numbers, and indices of linear operators, Uspehi Mat. Nauk 12 (1957), no. 2 (74), 43-118; English transl., Amer. Math. Soc. Transl. (2) 13 (1960), 185-264.

2. F. John, Plane waves and spherical means applied to partial differential equations, Interscience Publishers, New York, 1955.

3. L. Nirenberg, and H.F. Walker, The null-spaces of elliptic partial differential operators in R^n, J. Math. Anal. Appl. 42 (1973), 271-301.

4. H.F. Walker, On the null-spaces of first-order elliptic partial differential operators in R^n, Proc. Amer. Math. Soc. 30 (1971), 278-286.

5. H.F. Walker, A Fredholm theory for a class of first-order elliptic partial differential operators in R^n, Trans. Amer. Math. Soc. 165 (1972), 75-86.

6. H.F. Walker, On the null-spaces of elliptic partial differential operators in R^n, Trans. Amer. Math. Soc. 173 (1972), 263-275.

University of Houston
Houston, Texas.

J A DONALDSON

A Uniqueness Class for Two Improperly Posed Problems Arising in Mathematical Physics

ABSTRACT

We solve a uniqueness problem for two abstract Cauchy problems which arise as generlizations of concrete initial value problems in mathematical physics.

INTRODUCTION

Let B be a Banach space upon which there is defined a strongly continuous group $G_A(y)$ $(-\infty < y < \infty)$ of bounded linear operators. Furthermore, let A be the infinitesimal generator of $G_A(y)$ and let D(A) denote the domain of A.

Under the assumption that u(t) is such that for each fixed $t \geq 0$

$$||G_A(y)u(t)|| = O(e^{-|y|^{2-\delta}}), \quad \delta > 0,$$

we shall prove the uniqueness of the strict solution of each of the Cauchy problems

$$I_1 \quad \begin{cases} u'(t) = -A^2u(t), & t > 0 \\ u(0) = \emptyset, & \emptyset \in D(A^2); \end{cases}$$

and

$$I_2 \quad \begin{cases} u''(t) = A^4u(t), & t > 0 \\ u(0) = \emptyset, & \emptyset \in D(A^4) \\ u'(0) = 0, & \end{cases}$$

which are respectively, the initial value problems for the inverse "abstract" heat equation, and the backward "abstract" beam equation of the theory of elasticity. We reduce the Cauchy problem I_k to a Cauchy problem J_k for a partial differential equation in B. Then it is shown that the uniqueness of the solution of I_k follows from the uniqueness of the solution of J_k. Here and in the sequel the subscript "k" will take on the values 1 and 2.

PRELIMINARIES

Let $C([0,\infty);B)$ denote the space of functions which are continuous over $[0,\infty)$ with values in B. For $p > 1$, let F_p be the space of test functions $f(x)$ satisfying the inequality

$$|f(x)| \leq C_1 e^{-C_2|x|^p}$$

which can be continued analytically to the complex plane so that the resulting entire function $f(z)$ satisfies the inequality

$$|f(z)| \leq C_3 e^{C_4|z|^p}.$$

In F_p a multiplier is given by any entire function $g(z)$ of order p such that

$$|g(x)| \leq C_\epsilon e^{\epsilon|x|^p}$$

for any $\epsilon > 0$ where C_ϵ depends upon ϵ and g. By $\hat{F}_p$ we denote the space of functions consisting of Fourier transforms of functions from F_p. Gel'fand and Silov [2] have shown that $\hat{F}_p = F_{p'}$ where $1/p + 1/p' = 1$.

Elements of the space $T(F_p)$ of continuous linear operators from F_p into B will be called B-valued distributions. The Fourier transform $\hat{T}$ of a B-valued distribution T is an element of $T(\hat{F}_p)$ and is defined by

$$\hat{T}(\hat{f}) = T(f(-x))$$

for all $f \in F_p$. Here $\hat{f}$ denotes the Fourier transform of f. The Fourier transformation is a topological isomorphism between the spaces $T(F_p)$ and $T(\hat{F}_p)$.

Finally, $U_{2,A}$ will denote the sub-class of functions $u(t)$ in $C([0,\infty);B)$ such that for every fixed $t \geq 0$ one has

$$||G_A(y)u(t)|| = O(e^{|y|^{2-\delta}})$$

for any $\delta > 0$ as $|y| \to \infty$.

UNIQUENESS RESULTS

In this section we show that I_k can have at most one strict solution in the class of functions $U_{2,A}$. That is we prove

Theorem 1. The Cauchy problem I_k can have at most one strict solution in $U_{2,A}$.

Before establishing this theorem, we transform I_1 into an initial value problem involving a partial differential equation in B and study the question of uniqueness for this problem. Since u(t) is a strict solution of I_1, u(t) is in the domain of A^2 for each t. It follows that

$$G_A(y)A^2u(t) = A^2G_A(y)u(t) = d^2/dy^2\{G_A(y)u(t)\}.$$

Upon applying $G_A(y)$ to I_1 we obtain the Cauchy problem

$$J_1 \quad \begin{cases} W_t = -W_{yy} \\ W(y,0) = G_A(y)\emptyset \end{cases}$$

for the inverse heat equation in the Banach space B. Here we have set $W(y,t) = G_A(y)u(t)$. In a similar manner I_2 is transformed into

$$J_2 \quad \begin{cases} W_{tt} = W_{yyyy} \\ W(y,0) = G_A(y)\emptyset \\ W_t(y,0) = 0 \end{cases}$$

by $G_A(y)$. We establish now the following result.

Theorem 2. If the Cauchy problem I_k has a strict solution W(y,t) in the class of B-valued functions of the real variables y and t satisfying for each fixed $t \geq 0$ the relation

$$||W(y,t)|| = O(e^{|y|^{2-\delta}})$$

for any $\delta > 0$ as $|y| \to \infty$, then this solution is unique.

Proof

We consider the equation J_k as an equation relating to a B-valued distribution which depends upon t as a parameter and which with respect to y belongs to $T(F_{2-\delta})$. Applying the Fourier transformation to the equation and auxilliary conditions in J_k we obtain initial value problems

$$L_1 \qquad \begin{cases} \hat{W}_t(s,t) = s^2\hat{W}(s,t) \\ \hat{W}(s,t) = \hat{\Phi}(s) \end{cases}$$

and

$$L_2 \qquad \begin{cases} \hat{W}_{tt}(s,t) = s^4\hat{W}(s,t) \\ \hat{W}(s,0) = \hat{\Phi}(s) \\ \hat{W}_t(s,0) = 0 \end{cases}$$

in the space $T(\hat{F}_{2-\delta})$. Here, $\Phi(y) = G_A(y)\emptyset$.

Since there exists an isomorphism between $T(F_{2-\delta})$ and $T(F_{2-\delta})$, the question of uniqueness of the solution of J_k is equivalent to the question of uniqueness of the solution of L_k. Now $\hat{W}(s,t) = e^{s^2t}\hat{\Phi}(s)$ is the unique solution of L_1 and $\hat{W}(s,t) = \cosh(s^2t)\hat{\Phi}(s)$ is the unique solution of L_2, since the functions e^{s^2t} and $\cosh(s^2t)$ are multipliers for $\hat{F}_{2-\delta}$ and $T(\hat{F}_{2-\delta})$ for any $t \geq 0$. We prove this assertion for L_1. Now the function $e^{-s^2(t-t_0)}$ is a multiplier for $\hat{F}_{2-\delta}$ for any t, where $0 \leq t \leq t_0$. It follows that

$$e^{-s^2(t-t_0)}T(f) = T(e^{-s^2(t-t_0)}f)$$

for all f in $\hat{F}_{2-\delta}$. It is sufficient to show that with the initial condition $\hat{\Phi}(s) = 0$ the only solution of the equation in L_1 is the function $\hat{W}(s,t) \equiv 0$. Now fix $t_0 > 0$ and fix a test function $\Psi(s)$ in $\hat{F}_{2-\delta}$. Then consider $e^{-s^2(t-t_0)}\Psi(s)$, $0 \leq t \leq t_0$, and apply to it the B-distribution $\hat{W}(s,t)$. Differentiating with respect to t, we obtain

$$\begin{aligned}(d/dt)[\hat{W}(s,t)(e^{-s^2(t-t_0)}\Psi(s))] &= [d/dt(\hat{W}(s,t))](e^{-s^2(t-t_0)}\Psi(s)) \\ &\quad + \hat{W}(s,t)(d/dt(e^{-s^2(t-t_0)}\Psi(s)) \\ &=[(d/dt)(\hat{W}(s,t))](e^{-s^2(t-t_0)}\Psi(s)) \\ &\quad + \hat{W}(s,t)(-s^2e^{-s^2(t-t_0)}\Psi(s))\end{aligned}$$

$$= [s^2\hat{W}(s,t)](e^{-s^2(t-t_o)}\Psi(s))$$

$$- s^2\hat{W}(s,t)(e^{-s^2(t-t_o)}\Psi(s))$$

$$= 0.$$

It follows that

$$[\hat{W}(s,t)(e^{-s^2(t-t_o)}\Psi(s))]$$

is constant on $[0,t_o]$. Since $W(s,0) = 0$, this constant is the zero element of B. Setting $t=t_o$, we find that $\hat{W}(s,t_o)(\Psi(s))= 0$ and since Ψ is any element of $\hat{F}_{2-\delta}$ it follows that the B-distribution $\hat{W}(s,t_o) = 0$ for any t_o as was to be proved. The uniqueness of the solution of L_2 can be established in a similar way. Theorem 2 is now established. We now move to the proof of theorem 1.

<u>Proof</u>

The uniqueness of the solution I_k may be inferred from the uniqueness of the solution of J_k. We show that this is true for I_1 and remark that a similar analysis holds also for I_2. Thus, suppose that $u_1(t)$ and $u_2(t)$ are in $U_{2,A}$ and that they are two strict solutions of I_1. The difference $W(t)=u_1(t)-u_2(t)$ will satisfy the initial value problem

$$K \quad \begin{cases} Z'(t) = -A^2Z(t) \\ Z(0) = 0. \end{cases}$$

Upon applying the transformation $G_A(y)$ to K we obtain the Cauchy problem

$$K' \quad \begin{cases} W_t = -W_{yy} \\ W(y,0) = 0 \end{cases}$$

where $W(y,t) = G_A(y)Z(t)$. However, by theorem 2, K' has only the solution $W(y,t) = 0$ in the class of B-valued functions which satisfy for each fixed $t \geq 0$ the relation

$$||W(y,t)|| = O(e^{|y|^{2-\delta}})$$

as $|y| \to \infty$. By hypothesis $G_A(y)Z(t)$ is in this class, and consequently

$$G_A(y)Z(t) = W(y,t) = 0.$$

From this relation it follows that

$$Z(t) = u_1(t) - u_2(t) = 0.$$

Thus, $u_1(t) = u_2(t)$ as was to be established.

APPLICATIONS AND REMARKS

First we consider a few applications to partial differential equations. Consider

$$M \quad \begin{cases} u_t = -u_{xx} \\ u(x,o) = \phi(x) \end{cases}$$

where our space of data and solutions will be $L_p(R)$ for $p \geq 1$. Here $A = \partial/\partial x$. By theorem 1, if the solution of M exist in the class of functions $u(x,t)$ satisfying for each $t \geq 0$ the inequality

$$||u(x+y,t)|| \leq C_1 e^{C_2|y|^{2-\delta}}, \quad \delta > 0,$$

then this solution is unique. (Note: Here $G_A(y)u(x,t)=u(x+y,t)$). This is an extension to L_p spaces of a result established by Tikhonov [3].

For our second example we consider an equation with variable coefficients in x. Let $f(x)$ be bounded and continuously differentiable on $(-\infty,\infty)$. Further, let $|f(x)| \geq \sigma > 0$. We consider the Cauchy problem

$$N \quad \begin{cases} u_t = -[f(x)\partial/\partial x]^2 u(x,t) \\ u(x,0) = \phi(x) \end{cases}$$

where our space of data and solutions is L_∞ of the x-axis.

ow let

$$k(x) = \int_0^x \frac{dz}{f(z)}$$

hen upon setting $A = f(x)\partial/\partial x$, it is seen that A generates a roup whose action on $\phi(x)$ is given by $G_A(y)\phi(x)=\phi(k^{-1}(y+k(x)))$. y theorem 1, (N) has at most one solution in the class of unctions $u(x,t)$ satisfying for each $t \geq 0$, the relation

$$||G_A(y)u(x,t)|| = ||u(k^{-1}(y+k(x)),t)|| = O(e^{|y|^{2-\delta}}).$$

> 0, as $|y| \to\infty$.

We remark that this example does not readily lend itself to nalysis by the classical or the modern (Gel'fand-Silov-Schwartz) orm of the method of Fourier transforms, since the equation as variable coefficients involving x.

The uniqueness result of this paper comprise a small part of longer work on abstract Cauchy problems [1]. In [1] we nvestigate questions of uniqueness and of existence of solu- ions of higher order abstract Cauchy problems.

R. Hersh and I have used some of the results of [1] to study wo point boundary value problems for abstract operator ifferential equations. Specifically we have obtained results nvolving existence and uniqueness questions for the two point oundary problem for the "abstract" wave equation.

EFERENCES

. J.A. Donaldson, The Abstract Cauchy Problem and Banach Space Valued Distributions, Unpublished Manuscript, Howard University.

. I.M. Gel'fand and G.E. Silov, Fourier Transforms of Rapidly Increasing Functions and Problems of Uniqueness for Solutions of the Cauchy Problem, Uspehi Matem. Nauk (N.S.) 8, No. 6, 3-54 (1953).

. A. Tikhonov, Theoremes d'unicite pour L'equation de la chaleur, Mat. Sb. 42, (1935), 199-216.

Supported in part by NSF Grant No. GY - 11066.

Howard University
Washington, D.C. 20059.

STEINBERG

Some Unusual Ill Posed Cauchy Problems and Their Applications

In this lecture we shall discuss the initial value problem for certain partial differential equations which are not of classical type, namely not elliptic, parabolic, or strictly hyperbolic. What the equations have in common is that the solvability of the initial value problem can be studied using a particular technique usually referred to as the Ovcyannikov Theorem or the abstract Cauchy-Kowalewski technique. I will not discuss this technique but I will indicate the kinds of results that we can obtain from this technique.

First, we always obtain existence, uniqueness, and continuous dependence on the parameters of the problem, that is, the problems are "well posed". The difficulty here is that the topological spaces in which we obtain our solutions either have a very restrictive topology, for example, spaces of analytic functions, entire functions of some finite order, or subspaces of infinitely differentiable functions such as the Gevrey functions, or the spaces have a very weak topology, such as the dual spaces of the above spaces. Thus our problems will not be well posed in the usual sense but will be well posed in some generalized sense. We note that the classical Cauchy-Kowalewski Theorem is an example of such a result.

The above comments concern the so-called space variable. In contrast to this we need not restrict the data of our problem to depend analytically on the time variable and in fact, we may allow this dependence to be continuous or integrable. In general we obtain local solvability in the time variable, however, it is <u>always</u> possible to obtain global solvability in the time variable by placing additional <u>modest</u> assumptions on the data of the problem.

We also note that these techniques have been pushed far enough to handle equations that are of high order in the time derivatives, systems of equations, and nonlinear equations. Good references for studying these results are Treves [1] and Nirenberg [2].

The equations we will discuss do not resemble many of the equations discussed in the other lectures at this conference. In many of those problems the equations behave very nicely for time moving in one direction, such as the heat equation, while our equations behave badly for both time directions, such as for the initial value problem for the Laplace equation. Also, in many problems discussed at this conference the operators that appear in the equations are assumed to have certain symmetry properties, an assumption not valid for our equations. This, we believe, justifies using some novel techniques to discuss the basic properties of our equations.

Let us now look at some simple examples of the types of problems that have been studied using these techniques.

I. The Parametric Oscillator Equation.

The parametric oscillator equation is given by

$$\frac{\partial P}{\partial t} = -\frac{1}{2}\left(\frac{\partial^2 P}{\partial x^2} - \frac{\partial^2 P}{\partial y^2}\right) + x\frac{\partial P}{\partial x} - y\frac{\partial P}{\partial y}, \quad P = P(x,y,t)$$

$$P(x,y,0) = P_o(x,y)$$

for $(x,y,t) \in \mathbb{R}^2 \times [0,T]$. This equation belongs to a class of equations

$$\frac{\partial P}{\partial t} = Q\left(x, y, \frac{\partial}{\partial x}, \frac{\partial}{\partial y}\right)P$$

where Q is a quadratic expression in $x, y, \partial/\partial x, \partial/\partial y$ and where no symmetry assumptions are placed on Q. Note that the right hand side of the parametric oscillator equation has principal part which is the wave equation and consequently is not of classical type. For an extensive discussion of these equations see Steinberg and Treves [3] where there are also references to many discussions in the physics literature of equations of this type. However, we will make a few comments on the solutions of this problem. First, the solutions bear some resemblance to the solutions of the backwards heat equation (explicit solutions of this particular equation can be found). For the initial data $\exp(-\alpha(x^2+y^2))$ there exists a classical solution of our equation for a finite interval of time. If one allows the solutions to be an analytic functional, then there exists a global in time solution. This may at first seem inconvenient, but if one considers that the quantities of physical interest are the moments of the solution (not the values at a point) and that analytic functionals possess well behaved moments, then this type of solvability is appropriate to this situation. For an independent discussion of the moment problem see Miller and Steinberg [4].

The next reasonable case to discuss would be an equation of the form

$$\frac{\partial P}{\partial t} = F\left(x, y, \frac{\partial}{\partial x}, \frac{\partial}{\partial y}\right) P$$

where F is Cubic in its variables. Such an equation is Gordon's equation, see Klauder & Sudarskan [5], Steinberg [6], for which very little is known. This equation is a fully quantal model for a laser and contains six space variables instead of two.

II. Shot-Put Noise

The type of equations that appear in shot-put noise are of infinite order, for instance,

$$\frac{\partial f}{\partial t} = F\left(t, \frac{\partial}{\partial x}\right) f \quad , \quad f = f(x,t)$$

$$f(x,0) = f_o(x)$$

$$F(t,\xi) = \sum_{n=0}^{\infty} a_n(t)\xi^n .$$

In some elementary cases some insight can be gained into the solvability of these equations by the use of the Fourier transform. For a more extensive treatment see Steinberg [7]. In some applications the equations are truncated to second order and it is still an open question under what conditions this approximation is valid.

III. Cauchy-Kowalewski

We note that our techniques can be used to obtain the most general nonlinear form of the Cauchy-Kowalewski Theorem, that is, where the data is analytic in the space variables and only continuous in the time variable, see Nirenberg [8]. It appears that any existence and uniqueness results for initial value problems that can be obtained using power series techniques can be more easily obtained using Ovcyannikov techniques.

IV. The Cauchy - Goursat Problems

An elementary problem of this type is

$$\frac{\partial^2 f}{\partial t_1 \partial t_2} = P\left(x, t_1, t_2, \frac{\partial}{\partial x}\right) f \qquad , \; f = f(t_1, t_2, x)$$

$$f(0, t_2, x) = f_1(t_2, x) \qquad , \; t_2 \geq 0$$

$$f(t_1, 0, x) = f_2(t_1, x) \qquad , \; t_1 \geq 0$$

where P is some partial differential operator. For an extensive discussion of this type of problem see DuChateau [9].

V. Degenerate Equations or Equations With Regular Singular Points

A simple problem of this type is

$$t \frac{\partial f}{\partial t} + A\left(x, t, \frac{\partial}{\partial x}\right) f = g \qquad , \qquad f = f(x,t), g = g(x,t)$$

$$f(x,0) = ?$$

where A is a regular partial differential operator and g is a given function. For a discussion of higher order cases and a discussion of what initial data is to be specified see Baouendi and Gaoulauic [10].

VI. Hydrodynamics

Let us consider the following simple problem:

$$\frac{\partial f}{\partial t} = a\frac{\partial f}{\partial \theta} + b\frac{\partial f}{\partial r} + cf, \quad f = f(\theta,t)$$

$$f(\theta,0) = f_0(\theta)$$

here $f(\theta,t)$ and $f_0(\theta)$ are to be periodic of period 2π in θ and ,b,c are real constants. Here $\partial/\partial r$ is a shorthand notation 'or the following operator. Let $u(r,\theta,t)$ be harmonic in the .nit disk with $u(1,\theta,t) = f(\theta,t)$ and then set

$$\frac{\partial f(\theta,t)}{\partial r} = \frac{\partial u(r,\theta,t)}{\partial r}\bigg|_{r=1}.$$

'sing the Poisson kernal one can find an explicit expression 'or $\partial f/\partial r$ and then one can easily verify that this operator is non-local pseudodifferential operator of order 1. It is also ossible to explicitly solve this problem using Fourier series nd then see that the problem is not classically well posed, hat is, the equation has a non-real characteristic.

This equation is a simple prototype for the equations des- ribing the movement of a bubble in a fluid. The equation orresponds to the boundary condition on the surface of the ubble in Lagrangian coordinates. For further discussions of hese problems see Ovcyannikov [11]. We note that there are any problems in hydrodynamics to which these techniques should pply.

'II. Representation of Lie Groups

Ve simply remark here that there is a close relationship between ur techniques and the theory of analytic vectors for Lie groups, see Steinberg [12] and Goodman [13].

'III. Hyperbolic Equations with multiple characteristics.

A simple example of this type of equation is

$$\frac{\partial^2 f}{\partial t^2} = \frac{\partial f}{\partial x} \quad , \quad f = f(x,t)$$

$$f(x,0) = f_0(x) \ , \quad \frac{\partial f}{\partial t}(x,0) = f_1(x)$$

Note that this problem corresponds to a physical problem where one measures the temperature and flux at one point for all time and then attempts to find the heat distribution. Also note that the principal symbol of our operator is $\partial^2/\partial t^2$ which is hyperbolic with respect to $t = 0$ and has a double characteristic and that the problem is easily solved using the Fourier transform. Using this explicit formula it is easily seen that the problem is not classically well posed. Our techniques can be used to study the initial value problem for higher order equations or systems of equations that are linear but may have multiple characteristics. The main hypothesis on the equations studied are that the characteristics be real and smooth function of the space and time variables. We also note that solvability is obtained spaces of Gevrey functions of their duals which can be seen to be a natural choice by looking at examples similar to the one above.

Equations of the type we study appear in magnetohydrodynamics. For a discussion of these problems see Steinberg [14]. It should be possible to extend these techniques to non-linear problems. Also we note that there is a relationship between our techniques and those of Beals [15] for studying problems with multiple characteristics.

REFERENCES

1. F. Treves, Ovcyannikov Theorem and hyperdifferential operators, Rio de Janeiro, I.M.P.A., 1969.

2. L. Nirenberg, An abstract form of the nonlinear Cauchy-Kowalewski Theorem, J. Diff. Geometry, 6 (1972) 561-576.

3. S. Steinberg andF. Treves, Pseudo-Fokker Planck equations and hyperdifferential operators, J. Diff. Eq., 8 (1970) 333-366.

4. M. Miller and S. Steinberg, The solution of moment equation associated with a partial differential equation with polynomial coefficients, J. Math. Phys., 14 (1973) 337-339.

5. J. Klauder & E. Sudarshan, Fundamentals of Quantum Optics, N.A. Benjamin, Inc., New York, 1968, Chapter 9, Section 4.

6. S. Steinberg, The Cauchy problem for differential equations of infinite order, H. Diff. Eq., 9 (1971) 591-607.

7. S. Steinberg, Infinite systems of ordinary differential equations with unbounded coefficients and moment problems, J. Math. Anal. Appl., 41 (1973) 685-694.

8. See [2].

9. P. DuChateau, The Cauchy-Goursat Problem, Memoirs A.M.S. 118, 1972.

10. M.S.Baouendi and C. Goulaouic, Cauchy Problems with Characteristic Initial Hypersurface, Comm. Pure Appl. Math., 26 (1973) 455-475.

11. L.V. Ovcyannikov, Singular operators in Banach spaces, Dokl. Akad. Nauk, SSSR, 163 (1965) 819-822; Soviet Math. Dokl. 6 (1965) 1025-1028.

12. S. Steinberg, Local groups and analytic vectors, submitted for publication.

13. R. Goodman, Differential operators of infinite order on a Lie group I. J. Math. Mach. 19 (1970) 879.

14. S. Steinberg. Existence and Uniqueness of Solutions of hyperbolic equations which are not necessarily strictly hyperbolic. To appear in the J. Diff. Eqns.

15. R. Beals. Semigroups and abstract Gevrey Spaces. J. Functional Anal. 10 (1972) 300-308.

University of New Mexico,
Albuquerque, New Mexico,
U.S.A.

H D MEYER

Half-plane Representations and Harmonic Continuation

ABSTRACT

Representations of boundary-integral type are presented for solutions of Laplace's equation on the half-plane. These are derived using related Lions-Magenes inspired results of Saylor [9], [11], and conformal mapping. Applications to numerical harmonic continuation are briefly discussed. Results carry over to R_+^n.

1. INTRODUCTION

The object of this paper is to present two representations of boundary-integral type for solutions of Laplace's equation on the half-plane. Also, we will discuss briefly how such representations can be used to numerically continue harmonic functions to all or part of the half-plane given approximate function values on particular finite subsets of points.

For the sake of simplicity, the functions involved in this paper will be taken as real-valued and results will be discussed only for the plane R^2. However, the discussion can be carried over to complex-valued functions and to finite dimensional spaces of higher dimension (also to solutions of more general elliptic type equations).

The main representation comes from using Lions-Magenes [6] inspired results of Saylor [9], [11] along with conformal mapping. It has the form

$$u(x) = \sum_{q=0}^{\infty} \int_{-\infty}^{\infty} \Delta_y^{(q)} \tilde{P}\,(x_1,x_2;y)\,d\mu_q(y), \qquad (1.1)$$

where $\tilde{P}$ is a function related to the Poisson kernal, $\Delta_y^{(q)}$ is the q-th iterate of an operator related to the Laplacian, the $\{\mu_q\}$ are Borel measures whose total variations satisfy a boundedness condition, $x = (x_1,x_2)$, and μ has boundary values in a (distributional) space to be specified later. The second representation to be presented will be an offshoot of the one above.

As concerns the applications segment of the paper and continuation, it is important to point out that the process of harmonic continuation is unstable. That this is so can be seen by considering the standard example which involves the harmonic functions

$$u_k(z) = \mathrm{Re}(\tfrac{z}{R})^k \quad , \ z = x_1 + ix_2, \tag{1.2}$$

which converge to zero on $\{|z|<R\}$ and diverge otherwise. The imposition of global a priori bounds eliminates the difficulties arising from instability.

Related boundary-integral results for both elliptic and parabolic problems are found in [5], [7], [9], [11]. Of particular interest with respect to continuation procedures here are the papers [1], [2], [3], [4], [8], [10].

2. A UNIT DISK RESULT

Before treating the half-plane situation, we present a representation for (distributional) solutions of Laplace's equation on the unit disk. For the sake of completeness, its proof will be indicated briefly. This result, in more generality, is due to Saylor [9], [11], and the reader is referred to his work for more detail since the discussion here is only sketchy. In the next section, we will use the unit disk representation and conformal mapping to obtain the half-plane results.

First we need some notation and some spaces. Let $\underset{\sim}{x} = (x_1,x_2)$ be a typical point in Euclidean R^2 space with $|\underset{\sim}{x}| = \sqrt{x_1^2 + x_2^2}$. Take $U = \{|x|<1\}$ to be the unit open disk, $\tilde{U}$ its closure, and $\partial U = \{|x|=1\}$ as the boundary. Also let $\Delta \equiv \frac{\partial^2}{\partial x_1^2} + \frac{\partial^2}{\partial x_2^2}$ be the Laplace operator.

We designate by $\mathcal{E}(U)$ the space of all infinitely differentiable functions on U and by $\mathcal{D}(U)$ those functions in $\mathcal{E}(U)$ with compact support. Both of these are equipped with the usual Schwartz topologies. $\mathcal{E}'(U)$ and $\mathcal{D}'(U)$ represent the strong duals of $\mathcal{E}(U)$ and $\mathcal{D}(U)$ and are the standard distribution spaces of Schwartz.

For K any compact subset, let

$$||\phi||_{C}(K) \equiv \sup_{x\in K} |\phi(x)| \tag{2.1}$$

Then, $C^\infty(K)$ is the space of all infinitely differentiable functions on K provided with the norm (2.1).

The next spaces, as will be seen shortly, are connected with the traces of the harmonic distributions. Define $H(\partial U)$ to be the space of all real analytic functions ϕ on ∂U. It can be described equivalently as the space of all functions $\phi \epsilon C^\infty(\partial U)$ for which positive constants A and B exist such that

$$||\Delta^{(q)}\phi||_{C(\partial U)} \leq A(2q)!B^q, \quad q = 0, 1, \ldots . \tag{2.2}$$

In (2.2), Δ is the Laplace -Beltrami operator and the superscript (q) indicates a q-th iterate. If (x_1, x_2) is replaced by polar coordinates (r,θ), then $\Delta = \frac{\partial^2}{\partial\theta^2}$ on ∂U. Occasionally, a subscript will be appended to operators such as Δ to denote the particular variable with respect to which the operator acts.

$H(\partial U)$ is topologized by writing it as

$$H(\partial U) = \bigcup_{B} H_B$$

and with respect to this giving it the inductive limit topology. In the above,

$$H_B = \{\phi \epsilon C^\infty(\partial U) \mid \ || \phi ||_B \equiv \sup_q \frac{||\Delta^{(q)}\phi||_{C(\partial U)}}{(2q)!B^q} < \infty\}.$$

Each H_B is equipped with the topology provided by the norm $||\ \ ||_B$ and is a Banach space.

Define $H'(\partial U)$ to be the dual of $H(\partial U)$. Elements h in $H'(\partial U)$ can be represented as

$$h(\phi) = \sum_{q=0}^{\infty} \int_{\partial U} \Delta^{(q)} \phi(x)\, d\mu_q(x) \tag{2.4}$$

for all $\phi \in H(\partial U)$, where the $\{\mu_q\}$ are Borel measures on ∂U satisfying

$$\mathrm{var}(\mu_q) \leq \frac{A(\epsilon)\epsilon^q}{(2q)!} \quad , \; q = 0,1,\ldots, \tag{2.5}$$

for any $\epsilon > 0$. In (2.5), $A(\epsilon)$ is a positive function which in general tends to infinity as ϵ approaches zero; $\mathrm{var}(\mu_q)$ is the total variation of μ_q.

The unit disk representation will hold for distributions in the space

$$S = \{u \in \mathcal{D}'(U) \,|\, \Delta u = 0\}, \tag{2.6}$$

where Δu is taken in the distributional sense and S is provided with the induced $\mathcal{D}'(U)$ topology. Note that members of S can also be regarded as classical solutions since the distributional solutions can be corrected on a set of measure zero to make this so.

For elements in S it is possible to define a linear trace operator γ which maps S onto $H'(\partial U)$ and which for functions $u \in C^\infty(\bar{U}) \cap S$ just gives the restriction of u to ∂U. Thus the functions in S can be considered as that class of harmonic functions whose boundary traces fall in $H'(\partial U)$.

Note that it can be shown that the problem

$$\begin{aligned} &\Delta u = 0, \; u \in \mathcal{D}'(U), \\ &\gamma(u) = u_0, \; u_0 \in H'(\partial U), \end{aligned} \tag{2.7}$$

has a unique solution. Also note that γ defines a topological and algebraic isomorphism between S and $H'(\partial U)$. The topologies involved are the induced weak-star of $\mathcal{D}'(U)$ for S and the weak-star for $H'(\partial U)$.

We are now ready for the unit disk representation.

Theorem 2.1 Let u be a harmonic solution belonging to S. Then it has representation

$$u(x) = \sum_{q=0}^{\infty} \int_0^{2\pi} \Delta_t^{(q)} P(r,\theta;t)\, d\mu_q(t), \tag{2.8}$$

where the $\{\mu_q\}$ are Borel measures on $[0, 2\pi]$ satisfying

$$\operatorname{var}(\mu_q) \leq \frac{A(\epsilon)\epsilon^q}{(2q)!}, \quad q = 0, 1, \ldots, \tag{2.9}$$

for any $\epsilon > 0$. In (2.8), $x = re^{i\theta}$,

$$P(r,\theta;t) = \frac{1}{2\pi} \frac{1-r^2}{1-2r\cos(\theta-t) + r^2}, \tag{2.10}$$

and $A(\epsilon)$ is a positive function which depends on the choice of u.

Sketch of Proof. It can be shown that with respect to the $\mathcal{E}(U)$ topology, $S \cap C^\infty(\bar{U})$ is dense in S. Thus, given any $u \in S$, it is possible to pick a sequence of harmonic functions $\{u_k\} \subset S \cap C^\infty(\bar{U})$ such that $u_k \to u$ in $\mathcal{E}(U)$. Since the Poisson integral formula holds for functions in $S \cap C^\infty(\bar{U})$, we can write

$$u_k(x) = \int_0^{2\pi} P(r,\theta;t)u_k(e^{it})dt = \langle P(r,\theta;\cdot), \gamma(u_k)\rangle, \tag{2.11}$$

where $x = re^{i\theta} \in U$ and $\langle , \rangle$ represents the duality between $H(\partial U)$ and $H'(\partial U)$. The bracketed part of (2.11) is legitimate since it is possible to show that $P(r,\theta;t) \in H(\partial U)$ when it is considered a function of $y = e^{it}$.

If we now let $k \to \infty$ and recall the continuity properties of γ, we have that

$$u(x) = \langle P(r,\theta;\cdot), \gamma(u)\rangle. \tag{2.12}$$

Recalling (2.4), reconsidered here in terms of the polar coordinate variable t instead of x, (2.8) then follows.

3. HALF-PLANE REPRESENTATIONS

We turn now to a study of the half-plane results. Let $v = (v_1, v_2)$ represent a point in the Euclidean plane R^2_v and $x = (x_1, x_2)$ a point in the plane R^2_x. Polar coordinates for v will be (r,θ). U, $\bar{U}$, and ∂U will be the same as before and are taken as subsets of R^2_v. Let $R^2_+ = \{x \in R^2_x | x_2 > 0\}$ with $\bar{R}^2_+$ its closure and $\partial R^2_+ = \{x \in R^2_+ | x_2 = 0\}$ its boundary.

We will be using the transformation

$$x_1 = \frac{2v_2}{(1+v_1)^2+v_2^2}, \qquad x_2 = \frac{1-(v_1^2+v_2^2)}{(1+v_1)^2+v_2^2}, \tag{3.1}$$

with inverse

$$v_1 = \frac{1-(x_1^2+x_2^2)}{x_1^2+(1+x_2)^2}, \qquad v_2 = \frac{2x_1}{x_1^2+(1+x_2)^2}, \tag{3.2}$$

which maps $\bar{U}$ 1-1 onto $\bar{R}^2_+$. It takes $\{|v| = 1\}$ onto ∂R^2_+ and maps concentric circles $\{|v| = r<1\}$ onto circles C_r in R^2_+. The circles C_r are symmetric with respect to the x_2-axis and C_{r_1} falls inside C_{r_2} if $r_1<r_2$.

The derivatives

$$D_1 \equiv -x_1(1+x_2)\frac{\partial}{\partial x_1} + \frac{1}{2}\,[x_1^2-(1+x_2)^2]\,\frac{\partial}{\partial x_2}, \tag{3.3}$$

$$D_2 \equiv -\frac{1}{2}[x_1^2-(1+x_2)^2]\frac{\partial}{\partial x_1} - x_1(1+x_2)\,\frac{\partial}{\partial x_2}, \tag{3.4}$$

$$D_3 \equiv (1+x_1^2)\,\frac{\partial}{\partial x_1} \text{ (operating on } \partial R^2_+), \tag{3.5}$$

$$\Delta \equiv D_3^2 = [(1+x_1^2)\,\frac{\partial}{\partial x_1}]^2 \text{ (operating on } \partial R^2_+), \tag{3.6}$$

will be relevant to our discussion. They come from writing $\frac{\partial}{\partial v_1}$, $\frac{\partial}{\partial v_2}$, $\frac{\partial}{\partial \theta}\big|_{r=1}$, and $\Delta|_{r=1}$, respectively, in terms of R^2_x variables. Taking $D = (D_1, D_2)$, differentiation involving components of D will be referred to as D-differentiation.

Using the above, we can define spaces which are analogs of the spaces encountered in the last section. Let $\mathcal{D}(R^2_+)$ be the space of infinitely D-differential functions having compact support on R^2_+. To topologize it, let K be any compact subset of R^2_+ and let $C^\infty_0(K)$ consist of those functions in $\mathcal{D}(R^2_+)$ having their support in K. Topologize $C^\infty_0(K)$ by means of the semi-norms

$$\|\phi\|_{m,K} \equiv \sum_{|\ell|\leq m} \sup_{x\in K} |D^\ell \phi(x)|, \tag{3.7}$$

where $\ell = (\ell_1, \ell_2)$ is a multi-index, ℓ_1 and ℓ_2 are non-negative integers, $|\ell| = \ell_1 + \ell_2$, and $D^\ell = D_1^{\ell_1} D_2^{\ell_2}$. Then represent $\mathcal{D}(R^2_+)$ as $\bigcup_K C^\infty_0(K)$ (the sets K becoming increasingly larger) and equip it with the inductive limit topology. Let $\mathcal{D}'(R^2_+)$ be the strong dual of $\mathcal{D}(R^2_+)$.

Next, let $H(\partial R^2_+)$ be the space of all infinitely D_3-differentiable functions ϕ on ∂R^2_+ for which positive constants A and B exist such that

$$\|\Delta^{(q)}\phi\|_{C(\partial R^2_+)} \leq A(2q)!B^q, \quad q = 0, 1, \ldots . \tag{3.8}$$

In (3.8), $\| \quad \|_{C(\partial R^2_+)}$ is defined the same as $\| \quad \|_{C(K)}$.

If

$$H_B = \{\phi \in H(\partial R^2_+) \mid |||\phi|||_B \equiv \sup_q \frac{\|\Delta^{(q)}\phi\|_{C(\partial R^2_+)}}{(2q)!\, B^q} < \infty\}, \tag{3.9}$$

then $|||\quad|||_B$ is a norm H_B is a Banach space, and we specify the inductive limit topology for $H(\partial R^2_+) = \bigcup_B H_B(\partial R^2_+)$. Let $H'(\partial R^2_+)$ be the strong dual of $H(\partial R^2_+)$.

Paralleling the space S, the half-plane representations will hold for distributions in the space,

$$S = \{u \in \mathcal{D}'(R^2_+) \mid \Delta u = 0\}, \tag{3.10}$$

where just as in the last section, these can be considered as classical solutions. S is equipped with the induced $\mathcal{D}'(R^2_+)$

topology.

Again, it is possible to define a linear trace operator. Transferring γ from R_v^2 to R_x^2, we obtain an operator γ which maps S onto $H'(\partial R_+^2)$ and which gives as its map the restriction of u to ∂R_+^2 for functions $u \in C^\infty(\bar{R}_+^2) \cap S$. Functions in S can be considered as those harmonic functions whose traces fall in $H'(\partial R_+^2)$.

Because of (2.7) and the correspondence set up between U and R_+^2, the problem,

$$\Delta u = 0, \quad u \in \mathscr{D}'(R_+^2), \qquad\qquad (3.11)$$
$$\gamma(u) = u_0, \quad u_0 \in H'(\partial R_+^2)$$

has a unique solution. Also, γ defines a topological and algebraic isomorphism between S and $H'(\partial R_+^2)$, where the topologies are the induced weak-star of $\mathscr{D}'(R_+^2)$ for S and the weak-star for $H'(\partial R_+^2)$.

With the background above, we now can present the main representation.

Theorem 3.1. Let u be harmonic on R_+^2 and belong to S (with boundary trace in $H'(\partial R_+^2)$). Then it has representation

$$u(x) = \sum_{q=0}^{\infty} \int_{-\infty}^{\infty} \Delta_y^{(q)} \tilde{P}(x,y)\, d\mu_q(y), \qquad (3.12)$$

where the $\{\mu_q\}$ are Borel measures on $(-\infty,\infty)$ satisfying

$$\mathrm{var}(\mu_q) \le \frac{A(\epsilon)\,\epsilon^q}{(2q)!}, \quad q = 0, 1, \ldots, \qquad (3.13)$$

for any $\epsilon > 0$. In (3.13),

$$\tilde{P}(x,y) = \frac{1}{\pi} \frac{x_2(1+y^2)}{(x_1-y)^2 + x_2^2}, \qquad (3.14)$$

and $A(\epsilon)$ is a positive function which depends on the choice of u.

Proof. This follows easily considering the discussion above and transformation (3.1). $\tilde{P}$ corresponds to P, $\Delta_y^{(q)}$ to $\Delta_t^{(q)}$ and the $\{\mu_q\}$ on $(-\infty,\infty)$ to $\{\mu_q\}$ on $[0, 2\pi]$, so that applying (3.1) to the representation given by Theorem 2.1 gives our result.

Consider now, $P(r,\theta;t)$ as given by (2.10), We have that $\Delta_t P = \frac{\partial^2 P}{\partial t^2} = \frac{\partial^2 P}{\partial \theta^2}$. Further, for $r < 1$, $\frac{\partial}{\partial\theta}$ transforms into

$$D_4 \equiv \left(\frac{1-x_1^2-x_2^2}{2}\right)\frac{\partial}{\partial x_1} + x_1 x_2 \frac{\partial}{\partial x_2} \quad . \tag{3.15}$$

Define

$$\Delta \equiv D_4^2 = \left[\left(\frac{1-x_1^2-x_2^2}{2}\right)\frac{\partial}{\partial x_1} + x_1 x_2 \frac{\partial}{\partial x_2}\right]^2. \tag{3.16}$$

If we replace Δ_t in (2.8) with $\frac{\partial^2 P}{\partial\theta^2}$ and map this over to R_+^2, we have

$$u(x) = \sum_{q=0}^{\infty} \int_{-\infty}^{\infty} \Delta_x^{(q)} \tilde{P}(x,y)\, d\mu_q. \tag{3.17}$$

If we next replace each of the measures $\{\mu_q\}$ by measures $\{\nu_q\}$ such that $d\nu_q = (1+y^2)d\mu_q$, we have the following corollary to Theorem 3.1 in which $\tilde{P}$ is replaced by the standard half-plane Poisson kernel.

Corollary 3.1. Let u be harmonic on R_+^2 and belong to S (with boundary trace in $H'(\partial R_+^2)$). Then it has representation

$$u(x) = \sum_{q=0}^{\infty} \int_{-\infty}^{\infty} \Delta_x^{(q)} P(x,y)\, d\nu_q(y), \tag{3.18}$$

where the $\{\nu_q\}$ are Borel measures on $(-\infty,\infty)$ such that $d\nu_q = (1+y^2)d\mu_q$, q=0,1..., where the $\{\mu_q\}$ are Borel measures satisfying (3.13). In (3.18),

$$P(x,y) = \frac{1}{\pi}\frac{x_2}{(x_1-y)^2+x_2^2} \; . \tag{3.19}$$

Before closing this section, it should be noted that solutions given by the usual Poisson integral formula clearly fall in the class of solutions covered by (3.12) and (3.18).

Also note that representation (2.8) extends to spheres in higher dimensional Euclidean R_v^n spaces. Thus the representations given above carry over to n-dimensions if one uses the transformation

$$
\begin{aligned}
x_i &= \frac{2v_{i+1}}{(1+v_1)^2+v_2^2+\ldots+v_n^2}, \quad i = 1, \ldots, n-1, \\
x_n &= \frac{1-(v_1^2+\ldots+v_n^2)}{(1+v_1)^2+v_2^2+\ldots+v_n^2},
\end{aligned}
\tag{3.20}
$$

which takes a sphere in R_v^n onto the upper half-plane $\bar{R}_+^n=\{x_n \geq 0\}$ of Euclidean R_x^n space. The applications in the next section also will carry over for the n-dimensional case.

4. APPLICATIONS

We conclude by discussing the application of the representations just derived to numerical harmonic continuation. Douglas, Cannon, Saylor, Meyer [1], [2], [3], [4], [8], [10], and others, as mentioned previously, have studied continuation procedures of a related nature.

Consider first the approximation of harmonic functions u on all of the half-plane assuming the following:

(1) $u(x) \in S$;

(2) $|u(x^j)-F(x^j)|<\epsilon$, where F is known on a finite set of points $\{x^j\} \subset C_\rho$, $0<\rho<1$ (the $\{x^j\}$ correspond to points $\{v^j\}$ on $\{|v| = \rho\}$);

(3) $A(\epsilon)$ (appearing in (3.13)) is known.

Note that $A(\epsilon)$ serves as the global bound mentioned in the Introduction as needed.

Using (3.12) and picking positive integer parameters Q and N, we have

$$u(x) = \sum_{q=0}^{Q} \sum_{k=1}^{N} \Delta_y^{(q)} \tilde{P}(x,y^k)\mu_q([y^{k-1/2},y^{k+1/2}]), \tag{4.1}$$

where points $(y^s,0)$ are images of the points

$$v^s = e^{i\frac{2\pi s}{N}}, \quad s = 0,\ 1/2,\ \ldots,\ N,\ N+1/2, \tag{4.2}$$

which are equally spaced on the unit circle in R_v^2. Explicitly, the $\{y^s\}$ are given by

$$y^s = \frac{\cos\frac{2\pi s}{N}}{1+\sin\frac{2\pi s}{N}}, \quad s = 0,\ 1/2,\ \ldots,\ N,\ N+1/2. \tag{4.3}$$

Each of the measures μ_q can be decomposed into positive and negative parts

$$\mu_q = \mu_q^+ - \mu_q^-, \tag{4.4}$$

where

$$\mu_q^+([y^{k-1/2}, y^{k+1/2}]),\ \mu_q^-([y^{k-1/2},y^{k+1/2}]) \geq 0, \tag{4.5}$$

$$\mathrm{var}(\mu_q^+),\ \mathrm{var}(\mu_q^-) \leq \frac{A(\epsilon)\epsilon^q}{(2q)!}. \tag{4.6}$$

Let us substitute the sequence of parameters $\{a_{k,q}, b_{k,q}\}$ for the parameters $\{\mu_q^+([y^{k-1/2}, y^{k+1/2}]),$ $\mu_q^-([y^{k-1/2}, y^{k+1/2}])\}$. Corresponding to (4.5) and (4.6), we require

$$a_{k,q},\ b_{k,q} \geq 0, \tag{4.7}$$

$$\sum_{k=1}^{N} a_{k,q},\ \sum_{k=1}^{N} b_{k,q} \leq A_q,\ q = 0,\ \ldots,\ Q \tag{4.8}$$

$$A_q \equiv A(\tfrac{1-R}{2})\ (\tfrac{1-R}{2})^q/(2q)!, \tag{4.9}$$

where R is selected so that $0<\rho<R<1$.

The above and (4.1) then suggest the following form for our approximation

$$U_{Q,N}(x,\{a_{k,q}, b_{k,q}\}) = \sum_{q=0}^{Q} \sum_{k=1}^{N} \Delta_y^{(q)}\tilde{P}(x,y^k)(a_{k,q}-b_{k,q}). \quad (4.10)$$

The actual approximation $U_{Q,N}(x)$ now comes from picking a set of $\{a_{k,q}, b_{k,q}\}$, not necessarily unique, such that

$$\max_j |F(x^j) - U_{Q,N}(x^j,\{a_{k,q}, b_{k,q}\})| \quad (4.11)$$

is minimized subject to the constraints (4.7) - (4.9). Determining the $\{a_{k,q}, b_{k,q}\}$ amounts to treating a standard linear programming problem.

Note that in evaluating the terms $\Delta_y^{(q)}\tilde{P}(x,y^k)$ used in the approximation, this is more easily accomplished by evaluating the equivalent quantities in R_v^2, than by proceeding directly.

The procedure presented just above, by virtue of (3.1), matches up with a corresponding unit disk harmonic continuation procedure based on (2.8). This latter procedure is essentially the same as still another procedure for the unit disk discussed by Douglas [4].

Douglas bounds the error in his approximation by estimating the error on $\{|v|=\rho\}$ and $\{|v|=R\}$ and then applying Hadamard's three-circle theorem. The estimate on $\{|v|=\rho\}$ comes from knowing $A(\frac{1-R}{2})$ and that for $\{|v|=R\}$ comes from the data.

It is an easy matter to modify Douglas' error results so that they hold for our unit disk case. If these results, in turn, are transferred to the half-plane, it is then easily seen that we have the following error estimate:

<u>Theorem 4.1</u> Let u satisfy (1) - (3) and $U_{Q,N}(x)$ be determined as discussed above. Let $0<\rho<R<1$. Then

$$|u(x) - U_{Q,N}(x)| \le 2^{1/2}[(1-\frac{\rho^2}{r^2})^{-1/2}+(1-\frac{r^2}{R^2})^{-1/2}] \times M_R^{1-\frac{\ln(\frac{r}{R})}{\ln(\frac{\rho}{R})}} [2\varepsilon+C_R(2^{-Q}+N^{-1}+\delta^2)]^{\frac{\ln(\frac{r}{R})}{\ln(\frac{\rho}{R})}} \quad (4.12)$$

for x on C_r, $\rho<r<R$, where the constants M_R and C_R depend on the choice of R and $\delta \equiv \sup\{\max_{|v|=\rho} |v-v^j|\}$.

From (4.12), we see that $U_{Q,N}(x)$ converges to u(x) as ϵ and δ tend to zero and N and Q to infinity. It should be pointed out that the constants in (4.12) become large as R tends to one. It also should be pointed out that a bound for $|u-U_{Q,N}|$ inside C_ρ is given by the estimate

$$|u-U_{Q,N}| \leq 2\epsilon + C_R(2^{-Q}+N^{-1}+\delta^2). \qquad (4.13)$$

This estimate bounds $|u-U_{Q,N}|$ on C_ρ (it is obtained in the course of deriving (4.12)) and holds for u inside C_ρ by virtue of the maximum principle.

A second minimax approximation based on (3.18) can be set up in a fashion paralleling that for $U_{Q,N}(x)$. The error estimate will be similar.

Note further, that our hypothesis that data be given on some C_ρ is not unreasonable. For many distributions of data points in R^2_+, it is possible to take the information from these and bound the error on some C_ρ. When this is so, we are in essence treating the same situation as above as far as the error estimate is concerned. When data is given this way, the approximation is determined by proceeding as before only using the new points $\{x^j\}$ in (4.11).

Observe also, that the continuation problem (1) - (3) is not a practical one for values of ρ close to one. In such instances, one would be measuring data at points tending toward infinity.

Before closing, we will briefly discuss continuation assuming data is given in a second way. We assume

(1') $u(x) \in S$;

(2') $|u(\zeta^j,Y) - F(\zeta^j,Y)| < \epsilon$, where F is known on a finite set of points $\{(\zeta^j,Y)\}$ on the line $\{x_2=Y\}$;

(3') $A(\epsilon)$ is known;

and wish to continue u to some region $\chi = \{|x_1|<x,\ \eta\le x_2\le Y\}\subset R^2_+$.

This is done by picking an approximation of form

$$U_{Q,N}(x,\{a_{k,q},b_{k,q}\}) = \sum_{q=0}^{Q} \sum_{|\xi^k - x_1|\le \tilde{X}} \Delta_{\underset{\sim}{y}}^{(q)} \tilde{P}(x,y^k)(a_{k,q}-b_{k,q}), \tag{4.14}$$

which is required to be a best minimax fit to the data, subject to the constraints (4.7) - (4.9). In (4.8), $\underset{\sim}{y}^k=(\xi^k,0)$. We pick the $\xi^k=k\Delta x$ to have equal spacing Δx and X is an appropriately chosen parameter. The R in (4.9) must be picked sufficiently close to one in the range $0<R<1$.

Error bounds can be found using techniques similar to those employed by Cannon and Douglas in [2]. An alternative approximation procedure based on (3.18) can also be formulated.

REFERENCES

1. J.R. Cannon, Error estimates for some unstable continuation problems, SIAM J., 12 (1964), pp. 270-284.

2. J.R. Cannon and J. Douglas, Jr., The approximation of harmonic and parabolic functions on half-spaces from interior data, Centro Internazionale Matematico Estivo, Numerical Analysis of Partial Differential Equations, (Ispra (Varese), Italy, 1969), Editore Cremonese Roma, 1969, pp. 193-230.

3. J. Douglas, Jr., Unstable Physical Problems and their Numerical Approximation, Ch. 11, Lecture Notes, Rice University, Houston, Texas.

4. J. Douglas, Jr., Approximate continuation of harmonic and parabolic functions, Numerical Solution of Partial Differential Equations, J.H. Bramble, ed., Academic Press, New York, 1966, pp. 353-364.

5. G. Johnson, Jr., Harmonic functions on the unit disc I, Illinois J. Math., 12 (1968), pp. 366-385.

6. J.L. Lions and E. Magenes, Problemes aux limites non homogenes. VII, Ann. Mat. Pura Appl., 63 (1963), pp. 201-224.

7. H.D. Meyer, A representation for a distributional solution of the heat equation, SIAM J. Math. Anal. 5 (1974).

8. H.D. Meyer, Analytic continuation of holomorphic functions on C^n, to appear.

9. R. Saylor, A generalized boundary-integral representation for solutions of elliptic partial differential equations, Doctoral thesis, Rice University, Houston, Texas, 1966.

10. R. Saylor, Numerical elliptic continuation, SIAM J. Numer. Anal., 4 (1967), pp. 575-581.

11. R. Saylor, Boundary values of solutions of elliptic equations Indiana Univ. Math. J., to appear.

This paper will also appear in SIAM J. Math. Anal.

Department of Mathematics,
Texas Tech University, Lubbock,
Texas 79409.

J R CANNON and R E EWING

The Locations and Strengths of Point Sources

ABSTRACT

A numerical scheme for determining the locations and strengths of point sources in 3-dimensional Euclidean space from the measurement of the potential on a portion of a plane is presented. Asymptotic convergence rates are derived. If δ denotes the sum of the absolute values of the errors in measurement and truncation, then the asymptotic error in the locations and strengths is estimated as $O\left(\left[\log\log\log\log \frac{1}{\delta} / \log\log\log \frac{1}{\delta}\right]^{b}\right)$ as δ tends to zero, where b is a positive constant less than one.

1. INTRODUCTION

We consider the problem of determining the locations and strengths of an unknown number of point sources in 3-dimensional Euclidean space E^3. This problem is related to the physical problem of locating dense masses in the earth from gravity data taken at the surface or in the air.

We assume that all of the point sources are located below the plane $z = z_0$ in an x,y,z-Cartesian-coordinate system on E^3. Let Ω denote a bounded domain in the relative topology of the plane $z = z_0$. We assume that at points $Q_i \in \Omega$, $l=1,2,\ldots$, we can measure the potential to within an accuracy ϵ, $\epsilon > 0$. Also we assume that the points Q_i, $i=1,2,\ldots$, are dense in Ω. Since the number of sources to be located is finite, the sources are contained in a compact set D below the plane $z = z_0$. Moreover, it is reasonable to assume an a priori knowledge of the compact set D and an a priori separation of the sources. Also, it is quite likely that a priori estimates of the masses and their number would be available. If P and Q are points in E^3, let $\rho(P,Q)$ denote the usual Euclidean distance between P and Q. The potential u at Q determined by point sources at P_j, $j = 1,\ldots,N$, with corresponding strengths $M_j > 0$, $j=1,\ldots,N$

is given by the formula

$$u(Q) = \sum_{j=1}^{N} \frac{M_j}{\rho(Q,P_j)} \quad . \tag{1.1}$$

When $Q \in \Omega$, we set

$$g(Q) = u(Q). \tag{1.2}$$

Consequently, our assumption of measurement can be stated as there exists a function $g^*(Q)$ defined on Ω such that

$$|g^*(Q_i) - g(Q_i)| < \epsilon \tag{1.3}$$

for all $i=1,2,3,\ldots$. Finally, the assumption of a priori estimates of the masses M_j, $j=1, \ldots,N$, their number, and the separation of the sources can be stated as follows:

There exist positive constants μ_1,μ_2, s, and a positive integer N', such that

$$0 < \mu_1 \leq M_j \leq \mu_2, \quad j = 1,\ldots, N, \tag{1.4}$$

that

$$\rho(P_j,P_k) \geq s > 0 \tag{1.5}$$

for all $j \neq k$, j, k = 1,...,N, and that

$$N \leq N'. \tag{1.6}$$

In summary, we shall describe a method for determining approximations to N, P_j, M_j, j=1,...,N from the data g^*, the formula (1.1), the conditions (1.4), (1.5), (1.6) and the condition that the $P_j \in D$.

In Section 2 we shall consider a numerical procedure for approximating N, P_j and M_j. The procedure involves a finite sequence of nonlinear programming problems. Section 3 contains a discussion of the analyticity of u and its approximations. Also preliminary estimates for the error analysis are discussed.

Section 4 continues the error analysis with a discussion of the asymptotic determination of N and the general asymptotic continuous dependence of the P_j and the M_j upon the data error and truncation error arising from the usage of only finitely many of the Q_i in the approximation procedure. Section 5 presents an estimate of the asymptotic dependence upon the error. The final section contains a discussion of the results and its relation to the physical problem of locating masses from gravity measurements.

2. THE APPROXIMATION PROCEDURE

For each positive integer K, let $\vec{R}_K$ denote a point in the Cartesian product $\mathcal{D}_K = \prod_{j=1}^{K} D_j$, where $D_j = D$, $j=1,\ldots,K$. Consequently, each component R_j of $\vec{R}_K$ represents some point in D. Let $S(R_j,\mu)$ denote the open ball in E^3 with center R_j and radius μ. We define a subset $\mathcal{D}_K^s$ of $\mathcal{D}_K$ in the following way. Let

$$\mathcal{D}_K^s = \{\vec{R}_K \in D_K : R_1 \in D, \tag{2.1}$$

$$R_2 \in D - S(R_1,\tfrac{s}{2}),\ldots,R_j \in D - \bigcup_{k=1}^{j-1} S(R_k,\tfrac{s}{2}),\ldots,R_K \in D - \bigcup_{k=1}^{K-1} S(R_k,\tfrac{s}{2})\}.$$

Clearly, $\mathcal{D}_K^s$ is compact. Let

$$\mathcal{B}_K = \{\vec{B}_K = (B_1,\ldots,B_K) \in E^K : \mu_1 \le B_j \le \mu_2,\ j=1,\ldots,K\} \tag{2.2}$$

Consider the function

$$G_K^{(n)}(\vec{B},\vec{R}) = \max_{1\le i\le n} |g^*(Q_i) - V_K(Q_i;\vec{B},\vec{R})| \tag{2.3}$$

over $\mathcal{B}_K \times \mathcal{D}_K^s$, where

$$V_K(Q;\vec{B},\vec{R}) = \sum_{j=1}^{K} B_j \frac{1}{\rho(Q,R_j)} \quad . \tag{2.4}$$

Since $G_K^{(n)}$ is bounded, the infimum exists. Set

$$\zeta_K^{(n)} = \inf_{\mathscr{B}_K \times \mathscr{D}_K^s} G_K^{(n)}. \tag{2.5}$$

Since $G_K^{(n)}$ is continuous over the compact set $\mathscr{B}_K \times \mathscr{D}_K^s$, there exists a point $(\vec{B}_K^{*(n)}, \vec{R}_K^{*(n)}) \in \mathscr{B}_K \times \mathscr{D}_K^s$ such that

$$\zeta_K^{(n)} = G_K^{(n)}(\vec{B}_K^{*(n)}, \vec{R}_K^{*(n)}). \tag{2.6}$$

Now set

$$\zeta^{(n)} = \min_{1 \leq K \leq N'} \zeta_K^{(n)}.$$

Clearly, there exists a least K for which $\zeta^{(n)}$ is attained. Denote this K by K_n^*. Hence we obtain $(K_n^*, \vec{B}_{K_n^*}^{*(n)}, \vec{R}_{K_n^*}^{*(n)})$ as an approximation to $(N, \vec{M}_N, \vec{P}_N)$, where the use of $\vec{M}_N$ as the strength vector associated with $\vec{P}_N$ as the location vector is clear from the discussion above. Note that associated with the approximate solution $(K^*, B_{K_n^*}^{*(n)}, R_{K_n^*}^{*(n)})$ is a potential

$$w(Q) = V_{K_n^*}(Q; \vec{B}_{K_n^*}^{*(n)}, \vec{R}_{K_n^*}^{*(n)}). \tag{2.7}$$

Since an asymptotic estimate in the error in the locations, strengths, and number of sources must be obtained from the difference of the potentials u and w, we must estimate u-w on Ω. In order to do this we note that since the exact number, locations and strengths are admissible parameters in the problem for $\zeta_N^{(n)}$, it follows trivially that

$$0 \leq \zeta^{(n)} \leq \zeta_N^{(n)} \leq \epsilon. \tag{2.8}$$

We assume now that the Q_i have been chosen such that

$$h_n = \sup_{Q \in \Omega} \min_{1 \leq i \leq n} \rho(Q, Q_i)$$

tends to zero as n tends to infinity. This is clearly the case from the asserted density of the Q_i; however, it is worth pointing out that there exist choices of the Q_i which will yield rates of convergence. For example,

$$h_n = O\left(\frac{1}{\sqrt{n}}\right)$$

for the Q_i selected on a uniform grid placed on the plane $z=z_0$.

Lemma 1. There exists a positive constant $C_1=C_1(N', \mu_2, D)$ such that

$$|u(Q) - w(Q)| \leq \epsilon + C_1 h_n \tag{2.9}$$

for all $Q \in \Omega$.

Remark 1. Note that as ϵ tends to zero and n tends to infinity w tends uniformly to u on Ω. In what follows it is convenient to define

$$\delta = \epsilon + C_1 h_n. \tag{2.10}$$

Remark 2. The minimization of $G_K^{(n)}$ is a non-linear programming problem which appears to be quite difficult. See [6].

3. PRELIMINARY ESTIMATES

The estimates presented here are related quite strongly to those derived in [4] by Cannon and Douglas. The reader will be referred to that article and others for the sake of conciseness here.

Since the potentials u and w are harmonic, they are analytic functions of their arguments x, y, and z except at the points $\vec{P}_N$ and $\vec{R}^{*(n)}_{K^*_n}$ respectively. Set

$$f = u-w, \tag{3.1}$$

and select a point Q_0 in Ω. Consider any ray in the plane $z=z_0$ originating from Q_0 and extending to infinity. Let α be a complex variable such that Re $\alpha > 0$ denotes the arc length along that ray originating from Q_0 with Re α increasing as the ray is traversed toward infinity. It is an elementary exercise to establish that in the complex α plane there exists a semi-infinite strip

$$\mathcal{S} = \{\alpha: \text{Re } \alpha > 0 \text{ , } |\text{Im } \alpha| < \tau\}, \tag{3.2}$$

where τ depends upon D (the location of the singularities of f), such that $f=f(\alpha)$ is an analytic function. In addition, it can be shown that there exist two positive constants $C_i=C_i(N',D,\mu_2)$, $i=1,2$ such that

$$|f(\alpha)| < C_1 \tag{3.3}$$

for all $\alpha \in \mathcal{S}$ and that

$$|f(\alpha)| < C_2\{\text{Re } \alpha\}^{-1} \tag{3.4}$$

for Re $\alpha > C_3$, where $C_3=C_3(D) > 0$. Recalling Ω is a domain in $z=z_0$, (2.9), and (2.10), we see that there exists a positive constant β such that for α real and satisfying $0 < \alpha < \beta$,

$$|f(\alpha)| < \delta. \tag{3.5}$$

Employing (3.3), (3.5) and a lemma of Lindelöf [1,2,3], there exists a positive constant $\tau_1 < \tau$ such that for all α satisfying Re $\alpha = 2^{-1}\beta$ and $|\text{Im } \alpha| \leq \tau_1$,

$$|f(\alpha)| < C_1{}^{2/3}\delta^{1/3}. \tag{3.6}$$

By an argument similar to that given in [4], we obtain the following result.

Lemma 2. There exists a positive constant $C_4=C_4(C_1,C_2,C_3)$ such that for all Q with $z \geq z_0$

$$|f(Q)| \leq C_4\{\log\log \tfrac{1}{\delta}\}^{-1}. \tag{3.6}$$

Proof. From the maximum principle for harmonic functions, it suffices to estimate $f(\alpha)$ for positive real α. Consider f in the rectangle

$$\mathcal{R} = \{\alpha : 2^{-1}\beta \le \text{Re } \alpha \le \infty, \quad |\text{Im } \alpha| \le \tau_1\}.$$

Elementary estimates of the harmonic function $\log C_1^{-1}|f|$ yields

$$|f(\tau_2)| \le C_5 \delta^{\theta(\tau_2)}, \tag{3.7}$$

where τ_2 is a positive real number and

$$\theta(\tau_2) = C_6 \exp\{-C_7 \tau_2\}. \tag{3.8}$$

But, from (3.4), we note that

$$|f(\tau_2)| \; C_2 \tau_2^{-1} \tag{3.9}$$

for $\tau_2 > C_3$. Consequently,

$$|f(\tau_2)| \le \min (C_5 \delta^{\theta(\tau_2)}, C_2 \tau_2^{-1}) \tag{3.10}$$

for $\tau_2 > C_3$. The result (3.6) follows from setting

$$C_5 \delta^{\theta(\tau_2)} = C_2 \tau_2^{-1} \tag{3.11}$$

and estimating τ_2 in terms of δ.

Having established an estimate for f in the half space $z \ge z_0$, we can now apply the lemma of Carleman [2,3,5] to extend the estimate into a spherical shell about any one of the singularities of f. The order of such estimates is

$$|f| = O(\{\log \log \tfrac{1}{\delta}\}^{-a}) \tag{3.12}$$

where $0 < a < 1$ and a depends only upon the distance from the singularity.

4. CONTINUOUS DEPENDENCE UPON THE DATA

We begin with the following result.

<u>Lemma 3</u>. For all δ sufficiently small, the approximation procedure of Section 2 will determine $K_n^* = N$.

Proof: The proof is by contradiction. We suppose the contrary. Then, there exists a sequence of δ_n tending to zero as n tends to infinity such that $K_n^* \neq N$, $n=1,2,3,\ldots$. Since K_n^* can attain at most a finite number of values, one such value is attained for an infinite subset of the positive integers n. It is no loss of generality to assume that the value of K_n^* is $N + 1$ on some subsequence $\{n_m\}$ of the sequence $\{n\}$. An application of the Bolzano-Weierstrass theorem yields a subsequence of $\{\vec{R}_{N+1}^{*(n_m)}\}$ and $\{\vec{B}_{N+1}^{*(n_m)}\}$ which converge respectively to a point $\vec{R}_{N+1}^*$ of $\mathcal{D}_{N+1}^s$ and a point $\vec{B}_{N+1}^*$ of $\mathcal{B}_{N+1}$. Clearly, the corresponding potentials w tend to a potential w_* which is equal to u on Ω. Consequently, $u-w_* \equiv 0$ for $z \geq z_o$ via the results of the previous section and $u-w_* \equiv 0$ may be extended arbitrarily close to any singularity of $u-w_*$. Since w_* contains $N+1$ source of strength at least μ_1, there exists one singularity of $u-w_*$ which is at least $2^{-1}s$ distance from the rest. Near that singularity $u-w_*$ is zero while elementary calculations show that $u-w_* < -1$.

Consider now the function V_N defined in (2.4). Thus,

$$V_N : \mathcal{B}_N \times \mathcal{D}_N^s \longrightarrow Y, \tag{4.1}$$

where $Y=V_N(\mathcal{B}_N \times \mathcal{D}_N^s)$ is a subset of the set of all continuous functions on Ω which is metrized by the uniform norm. The map is clearly onto Y and is one-to-one via the argument of the previous lemma. Since Y is Hausdorff in the relative topology generated by the uniform norm, it follows [7, p. 141] that V_N is a homeomorphism. Thus we have demonstrated the following result.

Lemma 4. For δ defined by (2.10),

$$\lim_{\delta\to 0} K_n^* = N, \tag{4.2}$$

$$\lim_{\delta\to 0} \vec{B}_{K_n^*}^{*(n)} = \vec{M}_N, \tag{4.3}$$

$$\lim_{\delta\to 0} \vec{R}^{*(n)}_{K_n^*} = \vec{P}_N. \tag{4.4}$$

5. ASYMPTOTIC ESTIMATES OF CONTINUOUS DEPENDENCE.

Set

$$\delta' = \{\log \log \tfrac{1}{\delta}\}^{-1}. \tag{5.0}$$

From the results of previous sections, it follows that for δ' sufficiently small there corresponds to each source in u a source in w. Moreover, the distance between corresponding locations must approach zero; and likewise the corresponding difference in strengths must approach zero. Regarding P_j as fixed, then for δ' sufficiently small all corresponding $R_j^{*(n)}$ must lie within $S(P_j, \frac{s}{8})$. Consequently, the singularities at P_j and $R_j^{*(n)}$ are bounded away from the remaining singularities in f=u-w by a fixed positive distance. Fix δ', P_j and $R_j^{*(n)}$, and let O_j denote the mid point of the line segment connecting P_j and $R_j^{*(n)}$. Note that P_j and $R_j^{*(n)}$ belong to $S(O_j, \frac{s}{8})$ and that the spherical shell $\mathcal{K} = S(O_j, \frac{s}{2}) - S(O_j, \frac{s}{4})$ contains no singularities of f. Moreover, from (3.12) it follows that there exists a positive constant $C_8 = C_8(N', \mu_2, D, s)$ and a constant a, $0 < a < 1$, such that

$$|f(Q)| \leq C_8 \delta'^{a} \tag{5.1}$$

for all $Q \in \mathcal{K}$. Also, we note that in $S(O_j, \frac{s}{2})$, the function f has the representation

$$f(Q) = \left\{\frac{M_j}{\rho(Q, P_j)} - \frac{B_j^{*(n)}}{\rho(Q, R_j^{*(n)})} + e_j(Q)\right\}, \tag{5.2}$$

where $e_j(Q)$ is a bounded harmonic function. Let $C_{10} = C(N', \mu_2, s)$ denote the bound for $e_j(Q)$. In other words

$$|e_j(Q)| \leq C_{10} \tag{5.3}$$

for all Q in $S(O_j, \frac{s}{2})$.

Let α denote the arc length along a ray originating at 0_j and contained in the plane passing through 0_j which is orthogonal to the line segment connecting P_j and $R_j^{*(n)}$. Consider the source

$$u_j(Q) = M_j\{\rho(Q,P_j)\}^{-1}. \tag{5.4}$$

Restricting Q to the ray described above, we obtain

$$f_j(\alpha) = u_j(Q). \tag{5.5}$$

We can extend α into the complex plane and obtain the analytic extension of f_j into the region

$$\mathcal{a} = \{\alpha\colon \operatorname{Re}\alpha > 0 \quad |\operatorname{Im}\alpha| < \zeta \operatorname{Re}\alpha\}, \tag{5.6}$$

where ζ is a real positive number and $0 < \zeta < 1$. It is an elementary computation, to show that

$$|f_j(\alpha)| < C_{11}\{\operatorname{Re}\alpha\}^{-1},\ \alpha \in \mathcal{a}, \tag{5.7}$$

where $C_{11}=C_{11}(\mu_2\zeta)$, Similarly, it follows from (5.2), (5.3), (5.4), (5.5) and (5.7) that there exists a positive constant $C_{12}=C_{12}(N',\mu_2,s,\zeta)$ such that

$$|f(\alpha)| \leq C_{12}\{\operatorname{Re}\alpha\}^{-1},\ \alpha \in \mathcal{a}, \tag{5.8}$$

where $f(\alpha)$ is the analytic extension of $f(Q)$ off of the ray into the complex α-plane.

For $\alpha \in \mathcal{a}$ and α real, we obtain from an application of Carleman's lemma [2,3,5],

$$|f(\alpha)| \leq C_{12}\alpha^{-1}(\delta')^{\theta_1(\alpha)}, \tag{5.9}$$

where

$$\theta_1(\alpha) = C_{13}\alpha^{C_{14}}, \tag{5.10}$$

$C_{13}=C_{13}(\mathcal{a},s)$ and $C_{14}=C_{14}(\mathcal{a},s)$. Now, for Q on the ray at a distance α from O_j, we have

$$\rho(Q,P_j) = \rho(Q,R_j^{*(n)}) = \sqrt{\alpha^2 + \nu^2} \quad , \tag{5.11}$$

where

$$\nu = \rho(O_j,P_j) = \rho(O_j,R_j^{*(n)}). \tag{5.12}$$

Consequently, (5.2) yields

$$|M_j - B_j^{*(n)}| \le \sqrt{\alpha^2 + \nu^2}\ |f(Q)| + \sqrt{\alpha^2 + \nu^2}\ |e_j(Q)|, \tag{5.13}$$

and it follows from (5.3) and (5.9) that

$$|M_j - B_j^{*(n)}| \le C_{12}\alpha^{-1}\ \sqrt{\alpha^2 + \nu^2}(\delta')^{\theta_1(\alpha)} + \sqrt{\alpha^2 + \nu^2}\ C_{10}. \tag{5.14}$$

Selecting $\alpha = \nu$, we obtain

$$|M_j - B_j^{*(n)}| \le C_{15}(\delta')^{\theta_1(\nu)} + C_{16}\nu, \tag{5.15}$$

where $C_{15}=C_{15}(N',\mu_2,s,\zeta)$ and $C_{16}=C_{16}(N',\mu_2,s)$.

Consider now a point Q^* which is located at a distance $\nu + \gamma$, $\gamma > 0$, from O_j on the ray containing P_j and originating at O_j. A similar application of the lemma of Carleman [2,3,5] yields

$$|f(Q^*)| \le C_{17}\gamma^{-1}(\delta')^{\theta_2(\gamma)}, \tag{5.16}$$

where $C_{17}=C_{17}(N',\mu_2,s)$ and

$$\theta_2(\gamma) = C_{18}\gamma^{C_{19}},$$

$$C_{18} = C_{18}(\mathcal{Q},s) \text{ and } C_{19}=C_{19}(\mathcal{Q},s). \tag{5.17}$$

From the triangle inequality and (5.2), we obtain

$$\begin{aligned} M_j \left|\frac{1}{\rho(Q^*,P_j)} - \frac{1}{\rho(Q^*,R_j^{*(n)})}\right| &\le |f(Q^*)| \\ &+ |e_j(Q^*)| + \frac{1}{\rho(Q^*,R_j^{*(n)})}\ |M_j - B_j|. \end{aligned} \tag{5.18}$$

Since

$$\rho(Q^*, P_j) = \gamma \text{ and } \rho(Q^*, R_j^{*(n)}) = 2\nu + \gamma, \tag{5.19}$$

it follows from (5.2), (5.3), (5.15), and (5.16), that

$$\begin{aligned} \frac{2\nu M_j}{(2\nu+\gamma)\gamma} &\le C_{17}\gamma^{-1}(\delta')^{\theta_2(\gamma)} + C_{10} \\ &+ (2\nu+\gamma)^{-1}C_{15}(\delta')^{\theta_1(\nu)} + (2\nu+\gamma)^{-1}C_{16}\nu. \end{aligned} \tag{5.20}$$

Recalling that $M_j \ge \mu_1$, we obtain

$$\begin{aligned} 2\nu &\le C_{17}(2\nu+\gamma)(\delta')^{\theta_2(\gamma)}\mu_1^{-1} + C_{10}(2\nu+\gamma)\gamma\,\mu_1^{-1} \\ &+ \gamma C_{15}(\delta')^{\theta_1(\nu)}\mu_1^{-1} + \gamma C_{16}\nu\,\mu_1^{-1} \\ &\le C_{20}(\delta')^{\theta_2(\gamma)} + C_{21}\gamma \end{aligned} \tag{5.21}$$

since ν and γ are bounded by s and δ' may be assumed to be less than one.

Since the estimate (5.21) holds for all positive γ, we would like to estimate the minimum of the right-hand side. Setting

$$g(\gamma) = C_{20}(\delta')^{C_{18}\gamma^{C_{19}}} + C_{21}\gamma,$$

we obtain

$$g'(\gamma) = C_{20}(\delta')^{C_{18}\gamma^{C_{19}}}(\log \delta')C_{18}C_{19}\,\gamma^{C_{19}-1} + C_{21}.$$

Considering $g'(\gamma) = 0$, we obtain

$$\gamma = \left[\left\{\frac{1}{C_{18}}\,\frac{\log\log\frac{1}{\delta'}+\ldots}{\log\frac{1}{\delta}}\right\}^{\frac{1}{C_{19}}}\right], \tag{5.22}$$

where the +... signifies some additional terms involving γ. This motivates the choice

$$\gamma = \left\{ \frac{1}{C_{18}} \left[\frac{\log \log \frac{1}{\delta'}}{\log \frac{1}{\delta'}} \right] \right\}^{\frac{1}{C_{19}}} \tag{5.23}$$

which yields

$$C_{20}(\delta')^{\theta_2(\gamma)} = C_{20} \left[\log \frac{1}{\delta'} \right]^{-1} . \tag{5.24}$$

Since the expressions (5.23) and (5.24) represent the range in which the true order lies, slightly better estimates could be achieved through more complicated choices of γ. But, we do not feel that much is to be gained from such an effort.

Consequently, recalling (5.0), the expressions (5.23) and (5.24) together with (5.21) yield

$$\rho(P_j, R_j^*(n)) = O\left(\left[\frac{\log \log \log \log \frac{1}{\delta}}{\log \log \log \frac{1}{\delta}} \right]^b \right) \tag{5.25}$$

for some b, $0 < b < 1$, as δ tends to zero. Noting that C_{14} can be selected to agree with C_{19} in the geometrical set-up of the Carleman lemma, it follows from (5.25) with $b = C_{19}^{-1}$ and (5.15) that

$$|M_j - B_j^*(n)| = O\left(\left[\frac{\log \log \log \log \frac{1}{\delta}}{\log \log \log \frac{1}{\delta}} \right]^b \right) \tag{5.26}$$

We conclude with a summation of the asymptotic results.

Theorem. The asymptotic error in the locations and strengths of point sources is $O\left(\left[\frac{\log \log \log \log \frac{1}{\delta}}{\log \log \log \frac{1}{\delta}} \right]^b \right)$, where δ is the sum of the measurement error and the truncation error arising from a finite number of measurement points, $b=b(s)$ satisfies $0<b<1$, and the constant in the Landau order symbol depends only upon N, μ_1, μ_2, D and s.

6. LOCATION OF SOURCES FROM GRAVITY MEASUREMENTS

Keith Miller has pointed out that when we consider a large force

$$\vec{F} = \vec{k} \tag{6.1}$$

and its vector addition to a perturbing force

$$\vec{p} = a\vec{i} + b\vec{j} + c\vec{k}\ , \tag{6.2}$$

where, the a, b, c are small relative to one, we see that

$$|\vec{F} + \vec{p}| = \sqrt{a^2 + b^2 + (1+c)^2} \tag{6.3}$$

$$= 1 + c + \text{Higher Order Terms.}$$

Consequently, only the perturbation component in the direction of the large force can be recovered from the measurement of the magnitude of the combination. Thus, from gravity measurements at the surface of the earth, we can expect to measure to some degree of accuracy only the partial derivative of the potential u with respect to z, where z is the coordinate vertical to the surface of the earth. The analysis of the sections 2, 3 and 4 applies with only minor modifications. The techniques of section 5 can be applied, but since the order of the singularity is wors we cannot really expect any improvement in the asymptotic continuity estimate of section 5. Consequently, we do not pursue these estimates.

ACKNOWLEDGEMENT

The authors wish to thank Keith Miller of the University of California at Berkeley for several stimulating conversations.

REFERENCES

1. Behnke, H. and F. Sommer, "Theorie der Analytischen Funktionen einer Komplexen Veranderlichen", Springer-Verlag, Berlin, 1962, p. 127-128.

2. Cannon, J.R., "A priori estimate for the continuation of the solution of the heat equation in the space variable", Ann. Mat. Pura Appl., 4(65) (1966), pp. 377-388.

3. Cannon, J.R., "A Cauchy problem for the heat equation", Ann. Mat. Pura Appl., 4(66) (1966), pp. 155-165.

4. Cannon, J.R., and Jim Douglas, Jr., "The approximation of harmonic and parabolic functions on half-spaces from interior data", Centro Internazionale Matematico Estivo, Numerical Analysis of Partial Differential Equations (Ispra (Varese), Italy, 1969), Editore Cremonese Roma, 1969.

5. Carleman, T., Fonctions Quasi Analytiques, Ganthier-Villars, Paris, 1926, pp. 3-5.

6. Daniel, J.W., The Approximate Minimization of Functionals, Prentice-Hall, Englewood Cliffs, New Jersey, 1971.

7. Kelley, J.L., General Topology, Van Nostrand, New York, 1955.

This research was supported in part by the National Science Foundation and Texas Tech University.

J.R. Cannon, The University of Texas at Austin, Texas 78712

R.E. Ewing, Oakland University, Rochester, Michigan 48063.

K MILLER

Efficient Numerical Methods for Backward Solution of Parabolic Problems with Variable Coefficients

Consider the problem of approximately continuing backward in time the solution $u(x,t)$ of a linear parabolic equation when given data g for u not at the initial time $t = 0$, but at a later time $t = T > 0$. That is,

$$u_t = \Sigma \; a_{ij} u_{x_i x_j} + \Sigma \; b_i u_{x_i} + cu \text{ in } \Omega \times [0,\infty)$$

$$u = 0 \text{ on the sides } \partial\Omega \times [0,\infty), \tag{1}$$

$$u(x,T) \approx g(x) \text{ given.}$$

Here Ω is a bounded domain in R^n with decently smooth boundary and the variable coefficients $a_{ij}(x,t)$, $b_i(x,t)$ and $c(x,t)$ remain uniformly parabolic and fairly smooth. This ill-posed problem is stabilized for times $t > 0$ (as is shown by log convexity type arguments for example, see [1] and [8]) by imposing a prescribed bound on the initial function $u(x,0)$. Writing (1) as an ordinary differential equation on the Hilbert space $L^2(\Omega)$, we have

$$u' = -L(t)u, \quad t \geq 0,$$

$$\|u(T)-g\| \leq \epsilon, \tag{2}$$

$$\|u(0)\| \leq E,$$

with $g \in L^2(\Omega)$ and ϵ, E all given. Let's write this in terms of the unknown initial function f and the evolution operator A which maps initial values $u(0) = f$ into final values $u(T) \equiv Af$:

$$\|Af-g\| \leq \epsilon,$$

$$\|Bf-0\| \leq E, \tag{3}$$

where B will be the identity operator except where otherwise stated.

Methods of partial eigenfunction expansion. (See Miller [5], 1964). Let $\phi_1, \phi_2, \ldots$ be a complete system of "eigenfunctions" which are simultaneously orthogonal with respect to both A and B; that is,

$$(A\phi_i, A\phi_j) = (A_j)^2 \delta_{ij},$$
$$(B\phi_i, B\phi_j) = (B_j)^2 \delta_{ij}. \tag{4}$$

For example, if B is the identity, then the weights $(A_j)^2$ will all be 1's and the functions ϕ_j and weights $(B_j)^2$ will be the orthonormal eigenfunctions and corresponding eigenvalues of the compact self-adjoint operator $A^T A$. Now expand f in terms of the $\phi_1, \phi_2, \ldots$ and expand the data g in terms of the $A\phi_1, A\phi_2, \ldots$;

$$f = \sum_1^\infty f_j \phi_j \ , \ g = \sum_1^\infty g_j A\phi_j. \tag{5}$$

Then (3) can be written in terms of the coefficients $f_1, f_2, \ldots$ and $g_1, g_2, \ldots$ as follows:

$$\|Af-g\|^2 = \|\Sigma(f_j-g_j)A\phi_j\|^2 = \Sigma|(f_j-g_j)A_j|^2 \le \epsilon^2$$
$$\|Bf-0\|^2 = \|\Sigma(f_j-0)B\phi_j\|^2 = \Sigma|(f_j-0)B_j|^2 \le E^2. \tag{6}$$

Assume now that the eigenfunctions $\phi_1, \phi_2, \ldots$ have been so ordered that the ratios A_j/B_j are nonincreasing with respect to j. Then truncate our expansion of g at exactly that order α just previous to where A_j/ϵ becomes $< B_j/E$.

Let $\xi^\alpha = \sum_1^\alpha g_j \phi_j$ denote that initial function obtained by this αth order eigenfunction expansion of the data function g. (i.e., we have set our jth order coefficient equal either g_j or 0 depending upon whether the weight A_j/ϵ or the weight B_j/E is the larger in (6).) Then it can be shown that

$$\|A\xi^{\alpha}-g\| \leq 2\epsilon$$
$$\|B\xi^{\alpha}-0\| \leq 2E, \tag{7}$$

and hence ξ^{α} is a "nearly-best-possible" approximation to f, in the sense that ξ^{α} satisfies nearly the same "fit to data" and "prescribed bound" as does f itself.

Now this is a pretty specialized method, because, except in certain special cases with certain geometries and constant coefficients where separation of variables can be applied, we just aren't usually given the eigenfunctions of A^TA, and they can be very difficult to obtain, However, where they're available, it's a <u>very</u> good method and should be used.

There's one feature of these methods, of great importance for applications, which I'd like to emphasize. Often the bound E is unknown; what do we do then? Well, I would advise (in what I termed Method 3) to compute the partial expansions $\xi^1, \xi^2, \ldots, \xi^N$, but stop at that <u>first</u> number N such that

$$\|A\xi^{N}-g\| \leq 2\epsilon \tag{8a}$$

Thus we see (from (7)) that $N \leq \alpha$ and so we have

$$\|B\xi^{N}\| \leq \|B\xi^{\alpha}\| \leq 2E \tag{8b}$$

even though E is unknown; i.e., we've obtained (nearly) the claimed fit to the data with the (nearly) smallest possible initial function.

Notice one more feature of Method 3; absolutely no use has been made of the particular B in the computation; thus we've obtained (nearly) the claimed fit to the data with the (nearly) smallest possible $\|B\xi^N\|$, <u>simultaneously</u> for every bound which can be written in terms of the eigenfunctions ϕ_j of A^TA (i.e., in form 6) with weights B_j such that the ratio A_j/B_j is non-decreasing. For example, <u>it may be that the initial time is not even known</u>, that the solution could be extended with a nice bound E to a time $t_0 << 0$; in that case (let's assume $L(t) = L$ is constant and self-adjoint) we might let $Bf = u(t_0) = e^{-t_0 L} f$ and (8) tells us that we extend nicely back to time t_0 even though t_0 is unknown. Or, in the case of $L(t) \equiv -\Delta$ then $\|Bf\|$ might denote $\|\Delta^k f\|$ which is equivalent to the 2kth order Sobolev norm of f. One might thus say from (8) that our method has obtained (nearly) the claimed fit to the data, with the (nearly) smoothest initial function that is possible.

Least squares methods. These and similar methods seem to have been discovered independently by several authors. My versions, found in 1965, but appearing (with numerical computations) only in 1970 [6], are explicitly based in the notion of a stabilizing "prescribed bound". The method of G. Backus 1970 [2] is essentially equivalent to what I termed Method 1, except within a probabilistic interpretation. I find that the recipe of Morozov 1967 [10] for choosing the parameter in Tihonov's method of regularization yields essentially what I termed Method 3. Previous uses of such "regularization" methods often involved somewhat hazy justification for the choice of the regularization parameter; I mention Bellman et al., 1966 [3], and Tihonov 1963 [11]. Cleve Moler, at this conference, has pointed out to me that such regularization methods have been used since the late 1930's to stabilize ill-conditioned matrix problems. See also Miller and Viano [9] for an exposition of both expansion and least square methods.

Notice that the unknown initial function f from (3) satisfies

$$\|Af-g\|^2 + (\epsilon/E)^2 \|Bf\|^2 \leq 2\epsilon^2. \quad (9)$$

Thus, let our approximation ξ be that initial function such that

$$\|A\xi-g\|^2 + (\epsilon/E)^2 \|B\xi\|^2 \text{ is minimized,} \quad (10)$$

i.e., the solution of the least squares equation

$$(A^TA+(\epsilon/E)^2B^TB)\xi = A^Tg. \tag{11}$$

Since ξ will also satisfy the claimed fit to data and prescribed bound of (3) (except for a factor of at most $\sqrt{2}$) this is also a nearly best possible method. Moreover, one can <u>compute</u> exactly the best possible error bound for any linear functional of the solution.

In practice the parabolic equation (1) will usually be replaced by a finite difference or finite element approximation on a discretization Ω_h of Ω, and A in (11) will then denote the matrix mapping the discrete initial function $\xi = u(0)$ into the discrete final solution $A\xi = u(T)$. If Ω is the square in two dimensions, for example, and Ω_h is a 50 x 50 discretization of Ω, then A will be a 2500 x 2500 matrix. The problem is that A, and hence A^TA in (11), is horribly <u>non sparse</u>. In this form, therefore, the method seems totally impractical for multi-dimensional problems.

<u>Stabilized quasi-reversibility</u>. See Miller 1972 [7]. Suppose L in (2) is self-adjoint, ≥ 0, and constant with respect to t, and for notational simplicity assume T = 1. Now perturb the equation (2) a bit, replacing L in (2) by F(L), where the function $F(\lambda)$ is $\approx \lambda$ for small λ but is bounded above for large λ. One then solves the perturbed equation backward,

$$\begin{aligned} v' &= -F(L)v \ , \ t \leq 1, \\ v(1) &= g, \end{aligned} \tag{12}$$

Then, if desired, solve the unperturbed equation forward with the initial values $\xi \equiv v(0)$ so obtained to yield $A\xi$.

<u>Problem</u>: Choose F such that we are assured, assuming (3), that

$$\|u(t)-v(t)\| \leq 2\epsilon^tE^{1-t} \ , \ \ 0 \leq t \leq 1, \tag{13}$$

which we know, from log convexity, to be the best possible error bound that could be expected (except perhaps for the factor of

2). Alternatively, choose F such that we are assured (nearly) the claimed fit to data and prescribed bound; i.e.,

$$\|A\xi-g\| \leq 2\epsilon, \quad \|I\xi\| \leq 2E. \tag{14}$$

It turns out to be sufficient (and almost necessary) for either criterion (13) or (14), that

$$H(\lambda) \leq F(\lambda) \leq G(\lambda), \ \lambda \geq 0, \text{ where} \tag{15}$$

$$G(\lambda) = \lambda\ , \quad 0 \leq \lambda < \alpha \equiv \log\ (E/\epsilon), \tag{16}$$

$$= \alpha, \ \alpha \leq \lambda,$$

$$H(\lambda) = \log\left(1-e^{(\lambda-\alpha)}\right), \quad 0 \leq \lambda < \alpha,$$

$$= -\infty\ , \ \alpha \leq \lambda.$$

The original function $F(\lambda) = \lambda - \delta\lambda^2$ of Lattes and Lions fails condition (15) greatly when ϵ/E is small.

Rather than first choosing F and then having to approximate the solution of (12), it is better to instead divide the [0,1] interval into n subintervals of duration $\Delta t = 1/n$ and then compute $v_j \equiv v(j\Delta t)$ exactly, starting with $v_n = g$, by

$$v_j = R(L)v_{j+1}\ , \quad j < n-1, n-2,\ldots,0, \tag{17}$$

where R may be chosen to be <u>any</u> rational function satisfying

$$e^{\Delta tH(\lambda)} \leq R(\lambda) \equiv e^{\Delta tF(\lambda)} \leq e^{\Delta tG(\lambda)}, \quad \lambda \leq 0. \tag{18}$$

Of course, one should first compute the complex roots $1/\xi_i$ and $1/\eta_i$ of the numerator and denominator of R and then carry out (17) one complex linear factor at a time,

$$v_j = k\left[\frac{I-\xi_r L}{I-\eta_r L}\left[\ \cdots\ \left[\frac{I-\xi_2 L}{I-\eta_2 L}\left[\frac{I-\xi_1 L}{I-\eta_1 L}\ v_{j+1}\right]\right.\right.\right., \tag{19}$$

which requires the solution of only <u>sparse</u> equations. Thus we get some very efficient methods. A. Ralston (in an abortive collaboration for whose collapse I hereby apologize) computed the R of orders $r = 1,2,3$ (and $\Delta t = 1$) which satisfy (18) with smallest possible ϵ/E. The third order R attained an ϵ/E of .0005.

The shortcoming of this method is that it doesn't extend well to very general A, and definitely not well to A(t).

The backward beam equation approach. See Buzbee and Carasso 1971 [4]. Once again, let A be self-adjoint, constant with respect to t, and let T = 1. Then $y(t) = e^{\alpha t}u(t)$, with $\alpha = \log (E/\epsilon)$, satisfies

$$\text{(a)} \quad y'' = (A-\alpha)^2 y \tag{20}$$

$$\text{(b)} \quad \|y(1)-e^{\alpha}g\| \leq E$$

$$\text{(c)} \quad \|y(0)-0\| \leq E$$

One then lets our approximation be v(t), where $w(t) = e^{\alpha t}v(t)$ is the solution of the two point boundary value problem for (20a) with $w(1) = e^{\alpha}g$ and $w(0) = 0$. Because the norm of any solution of (20a) must be convex with respect to t, one gets the best possible error bound $\epsilon^t E^{1-t}$ once again.

The shortcoming here is that we have to simultaneously solve for all time levels at once; it introduces one higher dimension to the storage and computational difficulties. For example, with a 50x50x50 grid in two space dimensions and the one t dimension, we would have an elliptic finite difference equation with 125,000 simultaneous unknowns. Hence, I think that it is not practical for problems with two or more space dimensions.

It does seem, however, that the method extends readily to variable A(t) (not with best possible stability) and perhaps even to nonlinear equations. It was talks with Al Carasso that challenged me to take a fresh look at the problem with variable coefficients. I would now like to propose the previously mentioned least squares methods, but with some modifications.

Least Squares, with reduced dimensionality for ξ. In a typical problem with smooth coefficients, most highly oscillatory initial functions ξ will damp out drastically by $t = T$, so the space of ξ we need to fool with should be quite small.
The following work is something that Paolo Manselli and I have been looking at and which will be published jointly elsewhere. Let $\phi_1, \phi_2, \ldots, \phi_N$ be an "approximate basis" for our space $L^2(\Omega)$ of initial functions ξ; let P denote the orthogonal projection onto their linear span; let $Q = I - P$ denote the projection onto their orthogonal complement, and suppose that

$$\|AQ\| \leq .1\epsilon/E. \tag{21}$$

Instead of (3) we have

$$\begin{aligned} &\|A(Pf)-g\| \leq \|A(P-I)f\| + \|Af-g\| \leq 1.1\epsilon, \\ &\|I(Pf)\| \leq \|If\| \leq E; \end{aligned} \tag{22}$$

i.e., the projection Pf satisfies nearly the same constraints as f itself; its high order part Qf just hardly enters into the "fit to the data" of f. Therefore, we can do the least squares approach of (9)-(11), but with ξ a linear combination of the $\phi_1, \ldots, \phi_N$ only, and with ϵ replaced everywhere by 1.1ϵ. The matrix in (11) is then only $N \times N$, and involves only computing the solutions $A\phi_1, \ldots, A\phi_N$ and their inner products.

Now, once we've guessed at a good "approximate basis" $\phi_1, \ldots, \phi_N$ it is simple enough to check computationally whether $\|AQ\|$ is sufficiently small, since $\|AQ\|^2$ is the spectral radius of QA^TAQ, which can be computed by the power method. This involves computing $[QA^TAQ]^n\phi$, where ϕ is any initial function (say ϕ_{N+1}) which has a nonzero component of the dominant eigenfunction of QA^TAQ. Recall that A^T is itself an evolution operator; it carries the initial values $u(0) = \xi$ into the final values $u(T) = A^T\xi$ for the parabolic equation $u' = -L^T(T-t)u$, $0 \leq t \leq T$.

Let's consider several possible choices for our approximate basis.

Suppose, for example, that Ω is the one-dimensional interval $[0,\pi]$. One good choice for $\phi_1,..,\phi_N$ might then be the Fourier basis sin 1x,...,sin Nx. If the coefficients a(x,t) (when extended periodically across the end points by symmetric reflection) are sufficiently smooth (say C^q), then the kernal function k for the evolution operator A,

$$A\,(x) = \int_0^{\pi} k(x,y)\phi(y)dy \qquad (23)$$

is known to be a smooth (C^{q+2}) function of y, and hence integration by parts q + 2 times in (23) yields that $\|A^{N+1}\| = O(N^{-q-2})$.

A better choice might be to let $\phi_1,...,\phi_N$; $\phi_{N+1},...$ be the eigenfunctions of L(0), since $u' = L(t)u \approx L(0)u$ for small t. If u(0) is a high order eigenfunction ϕ_{N+1} of L(0), then the solution ought to die out exponentially so fast (initially at an e^{-cN^2t} rate recall) that it becomes very small before L(t) can change much from L(0). (We wish to assume here that the coefficients are not necessarily C^∞, but merely C^2 or so). Manselli and I have the following loose conjecture: that $\|A\phi_{N+1}\| = O(e^{-cN})$ at least, instead of merely $O(N^{-q})$. We haven't been able to prove this yet, but we ought to be able to. We have a bit of computational experience with a test program and our intuition appears to be justified; the solution starting out in a high order eigencomponent (with respect to L(0)) doesn't diffuse very quickly into the low order components.*

The best possible choice of our orthonormal basis to make $\|A\phi_1\|, \|A\phi_2\|,...$ decrease as quickly as possible is the eigenfunctions of A^TA of course. With this in mind we could start with an initial set of functions $\psi_1,...,\psi_N$, then do a

*Note added several months later: our intuition let us down. We have developed a very interesting counterexample.

refinement of them, rotating span $(\psi_1,\ldots,\psi_N)$ approximately into span (first N eigenfunctions of A^TA) by the block power method, applying powers of A^TA. Manselli and I are writing a program to do this for examples on the interval $[0,\pi]$. Starting with the initial basis sin 1x,..., sin Nx, we hope to see a big improvement in $\|AQ\|$ (computed by the power method recall) with only a few iterates of this refinement process.

Or, if we really wanted to do a lot of preliminary refining, we could strip off the eigenfunctions $\phi_1,\phi_2,\ldots,\phi_N$ of A^TA one at a time by the power method until we reached N with $\|A\phi^{N+1}\| \leq \epsilon/E$. If we did this very accurately, then we would of course be in a position to apply the eigenfunction expansion method instead of least squares. I would not recommend attempting such accurate refinement.

Nonlinear parabolic equations In this case I would attempt to minimize

$$\|A(\xi)-g\|^2 + \left(\frac{\epsilon}{E}\right)^2 \|B\xi\|^2, \qquad (24)$$

over all linear combinations ξ of some good approximate basis (say the Fourier basis), where A now denotes the nonlinear evolution operator mapping initial values $u(0) \equiv f$ into final values $u(T) \equiv A(f)$.

REFERENCES

1. Agmon, S., Unicite et convexite dans les problemes differentiels, Seminar Univ. de Montreal, Les Presses d l'Univ. Montreal, 1966.

2. Backus, G., On inference from incomplete and inaccurate data I, Proceedings Nat'l Acad. Sci., 65 (1970), pp. 1-8.

3. Bellman, R., Kalaba, R., and Lockett, K., Numerical inversion of the Laplace Transform, American Elsevier, New York, 1966.

4. Buzbee, B., and Carasso, A., On the numerical computation of parabolic problems for preceding times, Technical Report No. 229, Dept. of Math. and Stat., U. Of New Mexico, Albuquerque, November, 1971.

5. Miller, K., Three circle theorems in partial differential equations and applications to improperly posed problems, Arch. Rat. Mech. Anal., 16(1964), pp. 126-154.

6. Miller, K., Least squares methods for ill-posed problems with a prescribed bound, SIAM J., Math. Anal., 1 (1970), pp. 52-74.

7. Miller, K., Stabilized quasi-reversibility and other nearly-best-possible methods for non-well-posed problems, Symp. on Non-well-posed Problems and Logarithmic convexit Springer Lecture Notes // 316 (1973), pp. 161-176. Also to appear in J. Diff. Eqns.

8. Miller, K., Nonunique continuation for certain ODE's in Hilbert space and for uniformly parabolic and elliptic equations in self-adjoint divergence form, Symp. on Non-well-posed Problems and Logarithmic Convexity, Springer Lecture Notes // 316 (1973), pp. 85-101.

9. Miller, K., and Viano, G.A., On the necessity of nearly-best-possible methods for analytic continuation of scattering data, J. Math. Phys., Vol. 14, No. 8 (August 1973) pp. 1037-1048.

10. Morozov, V.A., Choice of parameter for the solution of functional equations by the regularization method, Dokl. Akad. Nauk SSR Tom 175 (1967), No. 6, (Translated in Soviet Math. Dokl., Vol. 8 (1967), No. 4, pp. 1000-10003)

11. Tihonov, A.N., On the solution of ill-posed problems and the method of regularization, Dokl. Akad. Nauk SSR, 151 (1963), pp. 501-504.

This work was supported in part by NSF Grant Number GP-29369X.

Department of Mathematics
University of California
Berkeley.

H BREZIS and J A GOLDSTEIN

Liouville Theorems for Some Improperly Posed Problems

1. INTRODUCTION

Of concern are equations of the form

$$d^n u/dt^n = (-1)^n Au, \quad t \in \mathbb{R}. \tag{1}$$

Here u maps the real numbers $\mathbb{R}$ into a real or complex Hilbert space H, A is a maximal monotone operator (possibly nonlinear, multivalued, and time dependent), and n = 1 or 2. We shall show that under some additional hypotheses, if u is a solution of (1), then either u is constant (as an H-valued function of t) or else $\|u(t)\|$ grows at least linearly as t tends to $+\infty$ or $-\infty$.

When n = 1, equation (1) for $t \le 0$ is (with an extra assumption on A) an abstract version of a backward parabolic problem, and when n = 2, (1) is an abstract version of an elliptic problem. Thus the Cauchy problem for the class of equations (1) under consideration is in general improperly posed.

We give now an example (from [7]) which explains why we call as "Liouville theorems" results of the type described in the first paragraph. Let X be the real space $L^p(\mathbb{R})$, $2 \le p < \infty$. Let f be a positive measurable functions on $\mathbb{R}$ and let $g(v) = f(\int_{-\infty}^{\infty} |v(x)|^p dx)$. Let $Av = -g(v)d^2v/dx^2$ for $v \in X$ be a continuously differentiable function such that v' is absolutely continuous and $v'' \in X$. Let $u : \mathbb{R} \longrightarrow X$ satisfy

$$d^2u/dt^2 = Au \quad (t \in \mathbb{R}),$$

which is an abstract version of

$$\partial^2 u/\partial t^2 + g(u)\partial^2 u/\partial x^2 = 0, \tag{2}$$

which is elliptic. Set

$$F(t) = \| u(t) \|_p^p = \int_{-\infty}^{\infty} |u(t,x)|^p dx.$$

Then a direct computation shows that

$$F''(t) = \int_{-\infty}^{\infty} p(p-1)|u|^{p-2}\{g(u)|\frac{\partial u}{\partial x}|^2 + |\frac{\partial u}{\partial t}|^2\}dx, \tag{3}$$

whence F is convex. Consequently either F is unbounded, in which case

$$\max\{\liminf_{t\to-\infty} \| t^{-1}u(t)\|_p, \liminf_{t\to+\infty} \| t^{-1}u(t)\|_p\} > 0,$$

i.e. $\| u(t)\|_p$ grows at least linearly as $t \to \pm\infty$, or else F is constant in which case $F'' = 0$, and so by (3), $u(t,x) \equiv$ constant ($= 0$ since $u(t, \cdot) \in L^p(\mathbb{R})$, $p < \infty$). Now suppose $f(s) = 1$ for all $s \in \mathbb{R}$. Then u, being a solution of (2) with $g \equiv 1$, is a harmonic function defined on the entire $t - x$ plane. The limiting case ($p = \infty$) of this example is the result : If $\sup_{t\in\mathbb{R}} \| u(t, \cdot)\|_\infty < \infty$ (i.e. if $u(t,x)$ is uniformly bounded in $(t, x) \in \mathbb{R}^2$), then $u \equiv$ constant. This is the classical theorem of Liouville.

2. SECOND ORDER EQUATIONS (ELLIPTIC PROBLEMS).

For each $t \in \mathbb{R}$ let $A(t)$ be a (possibly multivalued) operator on H. By a solution of

$$u'' \in A(t)u \quad (' = d/dt) \tag{4}$$

on $\mathbb{R}$ we mean a strongly continuously differentiable function u having a locally strongly absolutely continuous derivative such that (4) holds a.e. In particular, $u(t) \in \mathcal{D}(A(t))$, the domain of $A(t)$, for a.e. $t \in \mathbb{R}$.

<u>Theorem 1</u>. Suppose that for each $t \in \mathbb{R}$, $A(t)$ is a monotone operator on a real or complex Hilbert space H, $0 \in \mathcal{D}(A(t))$, and $0 \in A(t)0$. If u is a solution of (4), then either $u \equiv$ constant or else

$$\max\{\liminf_{t\to-\infty} \|t^{-1}u(t)\|, \liminf_{t\to+\infty} \|t^{-1}u(t)\| \} > 0,$$

so that $\| u(t) \|$ grows at least linearly as t tends to $-\infty$ or $+\infty$.

Proof. This result is a slight variant of [7, Theorem 2]. For facts about monotone operators see [1]. Let $F(t) = \| u(t) \|^2$. A straightforward computation shows that F and F' are locally absolutely continuous and

$$F''(t) = 2\| u'(t) \| + 2\mathrm{Re}(u''(t), u(t)) \text{ a.e.}$$

But

$$\mathrm{Re}(u''(t), u(t)) \in \mathrm{Re}(A(t)u(t) - A(t)0, u(t)-0) \subset [0,\infty[$$

since A(t) is monotone, Consequently

$$F''(t) \geq 2\| u'(t) \|^2 \text{ a.e.,}$$

so F is convex. The theorem follows easily.

Remark 1. If $u \equiv$ constant $= u(0)$, then $u(0) \in N(A(t))$, the null space of A(t), for a.e.t. In particular, $u \equiv 0$ if we assume that each A(t) is injective.

Remark 2. We now indicate why the Cauchy problem for (4) is improperly posed (cf. [5]). Let A be a linear maximal monotone operator on a complex Hilbert space. Then $-A$ generates a (linear) contraction semigroup $\{T(t) : t \geq 0\}$. If the Cauchy problem for (4) is well-posed, then (see [4], [6]) A generates a semigroup $\{S(t) : t > 0\}$ analytic in the right half plane. Semigroup theory tells us that $S(t)(H) \subset \mathfrak{D}(A)$ and $T(t) = S(t)^{-1}$ for $t > 0$. Hence

$$H = S(t)T(t)(H) \subset S(t)(\mathfrak{D}(A)) \subset \mathfrak{D}(A) \subset H,$$

and so A is bounded by the closed graph theorem.

3. FIRST ORDER EQUATIONS (BACKWARD PARABOLIC EQUATIONS).

Theorem 2. Let B be a maximal monotone linear operator on a complex Hilbert space H and suppose B^2 is monotone. If u is a solution of

$$u' + Bu = 0 \quad (t \in \mathbb{R}), \tag{5}$$

then either $u \equiv$ constant or else

$$\liminf_{t \to -\infty} \| t^{-1} u(t) \| > 0.$$

<u>Proof</u>. For B a linear maximal monotone operator, B^2 is monotone if and only if B is sectorial with semi-angle $\pi/4$; this was recently shown by Showalter [10]. Thus $-B$ generates a semigroup $\{S(t): t \geq 0\}$ having the property that $S(t)(H) \subset \bigcap_{n=1}^{\infty} \mathcal{D}(B^n)$ for each $t > 0$ (cf. [8;pp.280, 489, 490]). For $-\infty < a < b < \infty$, any solution of (5) has the form

$$u(b) = S(b - a)u(a);$$

hence (i) $u(b) \in \mathcal{D}(B^2)$ for all real b, and (ii) any solution of (5) also satisfies

$$u'' = B^2 u \quad (t \in \mathbb{R}).$$

Moreover, $\| u(t) \|$ is a nonincreasing function of $t \in \mathbb{R}$. Theorem 2 now follows as a consequence of Theorem 1.

Now let H be a real Hilbert space and let $\phi: H \longrightarrow \mathbb{R} \cup \{+\infty\}$ be a proper ($\phi \not\equiv +\infty$) convex lower semi-continuous function. Let

$$D_\phi = \{u \in H: \phi(u) < \infty\}.$$

For $u \in D_\phi$ define

$$\partial\phi(u) = \{f \in H: \phi(v) - \phi(u) \geq (f, v-u) \text{ for all } v \in D_\phi\}.$$

$\partial\phi$ is called the sub-differential of ϕ. For a review of the basic facts about sub-differentials see [1], [2].

<u>Theorem 3</u>. Let ϕ be a proper convex lowersemicontinuous function on a real Hilbert space H. Let u be a solution of

$$u' + Au \ni 0 \quad (t \in \mathbb{R}) \tag{6}$$

where $A = \partial\phi$. Then either $u \equiv$ constant or else

$$\liminf_{t\to-\infty} \| t^{-1}u(t) \| > 0.$$

<u>Proof</u>. To say that u is a solution of (6) means that $u: \mathbb{R} \longrightarrow H$ is locally strongly absolutely continuous and (6) holds a.e.

Taking the inner product of (6) with $u'(t)$ yields

$$\| u'(t) \|^2 + (d/dt)\phi(u(t)) = 0. \tag{7}$$

(This is easily seen if ϕ is smooth. In the general case approximate ϕ by a sequence of smooth convex functions and obtain (7) by a passage to the limit; cf. [2].). Let $-\infty < a < b < \infty$. Multiply (7) by $t - a$ and integrate from $t = a$ to $t = b$, integrating the term involving ϕ by parts. The result is

$$\int_a^b (t-a) \| u'(t) \|^2 dt + (b-a)\phi(u(b)) - \int_a^b \phi(u(t))dt = 0. \tag{8}$$

Note that $u(b) = S(b - a)u(a)$ $(= \lim_{n\to\infty}(I + \frac{t}{n}\partial\phi)^{-n}u(a))$ where $\{S(t): t \geq 0\}$ is the contraction semigroup determined by $-A = -\partial\phi$ (cf. [1]). For $a > b$, $h > 0$,

$$\begin{aligned} \| u(b+h) - u(b) \| &= \| S(b-a)u(a+h) - S(b-a)u(a) \| \\ &\leq \| u(a+h) - u(a) \|, \end{aligned}$$

whence $\|u'(t)\|$ is a nonincreasing function of t. Thus

$$\begin{aligned} \tfrac{1}{2}(b-a)^2 \|u'(b)\|^2 &\leq \int_a^b (t-a) \| u'(t) \|^2 dt \\ &= -\int_a^b (\phi(u(b)) - \phi(u(t)))dt \text{ (by (8))} \\ &\leq -\int_a^b (-u'(t), u(b) - u(t))dt \end{aligned}$$

(since $u' \in -\partial\phi(u)$ and by the definition of $\partial\phi(u)$)

$$= -\frac{1}{2}\int_a^b \frac{d}{dt}\|u(b)-u(t)\|^2 dt = \frac{1}{2}\|u(b)-u(a)\|^2.$$

Consequently

$$\|u'(b)\| \leq \|u(b)-u(a)\|/(b-a) \tag{9}$$

whenever $-\infty < a < b < \infty$. If $\liminf_{t\to-\infty} \|t^{-1}u(t)\| > 0$ is false, there is a sequence $t_n \to -\infty$ such that $\|t_n^{-1}u(t_n)\| \to 0$. Fix $b \in \mathbb{R}$ and let $a = t_n$ in (9). Letting $n \to \infty$ yields $\|u'(b)\| = 0$, which holds for all real b, and so $u \equiv$ constant.

Remark 3. The idea of the proof of Theorem 3 can also be applied to Theorems 1 and 2. Thus, for instance, for an alternate approach to Theorem 1, take the inner product of $u''(t) \in A(t)u(t)$ with $u(t)$, etc., and use results from [3]. We omit the details.

4. SOME COMPLEMENTS AND GENERALIZATIONS

Theorem 4. Let B be a closable linear operator on a real or complex Hilbert space H and suppose B^2 is monotone. Let $u: \mathbb{R} \to (B^2)$ be a solution of

$$u' + Bu = 0 \quad (t \in \mathbb{R});$$

then either $u \equiv$ constant or else

$$\max\{\liminf_{t\to-\infty} \|t^{-1}u(t)\|, \liminf_{t\to+\infty} \|t^{-1}u(t)\|\} > 0.$$

Proof This follows, from Theorem 1 in the same way as does Theorem 2.

For example, taking B to be any symmetric operator, we recover a result due to Zaidman [11] and Levine [9]. See [7] and [9] for other extensions of this result.

Note that Theorems 2 and 3 are both extensions of the Zaidman-Levine result for the special case $B \geq 0$.

Another extension of the Zaidman-Levine (selfadjoint) result is obtained by letting $B = A_1 - A_2$ in Theorem 4 where

the linear operators A_1, A_2 satisfy the following two conditions:
(i) A_j is maximal monotone and A_j^2 is monotone, $j = 1, 2$.
(ii) If $\{S_j(t): t \geq 0\}$ is the semigroup generated by $-A_j$; then $S_j(t)$ is a normal operator and $S_1(t)S_2(s) = S_2(s)S_1(t)$ for $j = 1, 2$ and all $t, s \geq 0$. We omit the details of the proof.

Next we treat some first order equations with time dependent operators.

<u>Theorem 5</u>. For each $t \in \mathbb{R}$ let $B(t)$ be a linear operator on a real or complex Hilbert space H such that $B(t)^2$ is monotone. Let $D \subset \cap \{\mathcal{D}(B(t)^2): t \in \mathbb{R}\}$ and suppose $u: \mathbb{R} \to D$ is a solution of

$$u' + B(t)u = 0 \quad (t \in \mathbb{R}).$$

If $B(\cdot)f \in C^1(\mathbb{R},H)$ for each $f \in D$ and $-B'(t)|_D$ is monotone, then either $u \equiv$ constant or else

$$\max\{\liminf_{t\to-\infty} \|t^{-1}u(t)\|, \liminf_{t\to+\infty} \|t^{-1}u(t)\|\} > 0.$$

<u>Proof</u>. $u'(t) = -B(t)u(t)$ is locally absolutely continuous, and its derivative satisfies $u''(t) = -B'(t)u(t) + B(t)^2u(t)$, $t \in \mathbb{R}$. Let

$$F(t) = \|u(t)\|^2, \quad t \in \mathbb{R}.$$

Then, as in the proof of theorem 1, we get

$$\begin{aligned} F''(t) &= 2\|u'(t)\|^2 + 2\mathrm{Re}(u''(t), u(t)) \\ &= 2\|u'(t)\|^2 + 2\mathrm{Re}(\{B(t)^2 - B'(t)\}u(t), u(t)) \\ &\geq 2\|u'(t)\|^2 \geq 0, \end{aligned}$$

and the Theorem follows from the convexity of F.

Note that the above proof actually shows that the assumption that $B(t)^2$ and $-B'(t)|_D$ are both monotone for all $t \in \mathbb{R}$ can be weakened to: $(B(t)^2 - B'(t))|_D$ is monotone for each $t \in \mathbb{R}$.

As an example of a $B(t)$ satisfying the above hypothesis, let B_0 be a linear operator on H with B_0^2 monotone, and let $B(t) = b(t)B_0$ for $t \in \mathbb{R}$, where b is a continuously differentiable function on $\mathbb{R}$. Taking $D = \mathcal{D}(B_0^2)$, the above assumption becomes

$$b'(t) \operatorname{Re}(B_0 f, f) \leq b(t)^2 \operatorname{Re}(B_0^2 f, f) \tag{10}$$

for each $f \in D$. Sufficient for this is that B_0 is monotone and b is nonincreasing. Or, if $B_0 \geq I$, sufficient for (10) is that b satisfies the differential inequality

$$b(t)^2 - b'(t) \geq 0, \quad t \in \mathbb{R},$$

which gives a one-sided bound on the growth of b.

<u>Theorem 6</u>. Let ϕ be a nonnegative proper convex lower semicontinuous function on a real Hilbert space H. Let $A(t) = \alpha(t)\partial\phi$, where $\alpha: \mathbb{R} \to \mathbb{R}$ is an absolutely continuous nonnegative nonincreasing function such that

$$\lim_{a \to -\infty} \frac{1}{a^2} \int_a^0 \alpha(t)dt = 0. \tag{11}$$

Let u be a solution of

$$u' + A(t)u = 0 \quad (t \in \mathbb{R}). \tag{12}$$

Then either $u \equiv$ constant or else

$$\liminf_{t \to -\infty} \|t^{-1}u(t)\| > 0.$$

<u>Proof</u>. Note that

$$\frac{d}{dt}(\alpha(t)\phi(u(t))) \in \alpha'(t)\phi(u(t)) + \alpha(t)(\partial\phi(u(t)), u'(t)).$$

Taking the inner product of (12) with $u'(t)$ yields

$$\|u'(t)\|^2 + \frac{d}{dt}(\alpha(t)\phi(u(t))) - \alpha'(t)\phi(u(t)) = 0, \tag{13}$$

as in the proof of Theorem 3. Let $-\infty < a < b < \infty$. Multiply

(13) by t - a and integrate from t = a, to t = b, integrating the middle term on the left side by parts. The result is

$$0 = \int_a^b (t-a)\,\|u'(t)\|^2 dt + (b-a)\alpha(b)\phi(u(b)) - \int_a^b \alpha(t)\phi(u(t))dt$$

$$-\int_a^b (t-a)\alpha'(t)\phi(u(t))dt \equiv J_1 + J_2 + J_3 + J_4.$$

Now, $J_4 \geq 0$ since ϕ is nonnegative and α is nonincreasing. Moreover,

$$\begin{aligned} J_2 + J_3 &= \int_a^b (\alpha(b) - \alpha(t))\phi(u(b))dt \\ &\quad + \int_a^b \alpha(t)(\phi(u(b)) - \phi(u(t)))dt \\ &\geq - \int_a^b \alpha(t)\phi(u(b))dt \\ &\quad + \int_a^b (-u'(t),\ u(b) - u(t))dt \\ &\geq - \int_a^b \alpha(t)dt\phi(u(b)) - \tfrac{1}{2}\,\|u(b) - u(a))\,\|^2; \end{aligned}$$

the reasoning **is** similar to that used in the proof of Theorem 3. Also if $\{S(t)\colon t \geq 0\}$ is the semigroup determined by $-\partial\phi$, then u is given by

$$u(b) = S\left(\int_a^b \alpha(\tau)d\tau\right)u(a)$$

for $-\infty < a < b < \infty$. It follows that $\|u'(t)\|$ is a nonincreasing function of $t \in \mathbb{R}$, (cf. the proof of Theorem 3). Combining all these estimates gives

$$\tfrac{1}{2}(b-a)^2\,\|u(b) - u(a)\|^2 + \int_a^b \alpha(t)dt\phi(u(b)),$$

$$\leq \tfrac{1}{2}\,\|u(b) - u(a)\|^2 + \int_a^b \alpha(t)dt\phi(u(b)),$$

or

$$\|u'(b)\|^2 \leq \frac{\|u(b)-u(a)\|^2}{(b-a)^2} + \frac{2}{(b-a)^2}\int_a^b \alpha(t)dt\phi(u(b)).$$

With the aid of (11), the proof is completed as was the proof of Theorem 3.

A careful examination of the proof shows that we may drop the assumption $\phi \geq 0$ and (11) if instead we assume

$$\limsup_{a\to-\infty}(b-a)^{-2}\{\int_a^b(t-a)\alpha'(t)\phi(u(t))dt$$

$$-\int_a^b(\alpha(b)-\alpha(t))dt\phi(u(b))\}\leq 0.$$

In this way we obtain a theorem which subsumes Theorems 3 and 6.

REFERENCES

1. H. Brezis, Operateurs Maximaux Monotones et Semi-Groupes de Contractions dans les Espaces de Hilbert, North Holland, Amsterdam, 1973.

2. H. Brezis, Monotonicity methods in Hilbert spaces and some applications to nonlinear partial differential equations, in Contributions to Nonlinear Functional Analysis (ed. by E. Zarantonello), Academic, New York, 1972, pp. 101-156.

3. H. Brezis, Equations d'evolution du second ordre associees a des operateurs monotones, Israel J. Math. 12 (1972), pp. 51-60.

4. H.O. Fattorini, Ordinary differential equations in topological vector spaces, I, J. Differential Equations 5 (1969), pp. 72-105.

5. J.A. Goldstein, Some remarks on infinitesimal generators of analytic semigroups, Proc. Amer. Math. Soc. 22 (1969), pp. 91-93.

6. J.A. Goldstein, On a connection between first and second order differential equations in Banach spaces, J. Math. Analysis Appl. 30 (1970), pp. 246-251.

7. J.A. Goldstein and A. Lubin, On bounded solutions of nonlinear differential equations in Hilbert space, SIAM J. Math. Analysis 5 (1974), to appear.

8. T. Kato, Perturbation Theory for Linear Operators, Springer, New York, 1966.

9. H.A. Levine, on the uniqueness of bounded solutions to $u'(t) = A(t)u(t)$ and $u''(t) = A(t)u(t)$ in Hilbert space, SIAM J. Math. Analysis 4 (1973), pp. 250-259.

10. R. Showalter, The Final value problem for evolution equations, to appear.

11. S. Zaidman, Uniqueness of bounded solutions for some abstract differential equations, Ann. Univ. Ferrara, Sez. 7 (N.S.) 14 (1969), pp. 101-104.

Haim Brezis, Institut de
Mathematiques, Universite de
Paris VI, 4 Place Jussieu
Paris, 5^{e}, France.

Jerome A. Goldstein,
Department of Mathematics,
Tulane University, New Orleans,
Louisiana 70118, U.S.A.

R E SHOWALTER

Quasi-Reversibility of First and Second Order Parabolic Evolution Equations

INTRODUCTION

We consider the (possibly) improperly posed final value problem

$$u'(t) + Au(t) = 0, \quad 0 < t < T \tag{E}$$
$$u(T) = f,$$

where A is a maximal accretive (linear) operator in a complex Hilbert space H. When the numerical range of A lies in the sector of those complex numbers z with $|\arg(z)| \leq \pi/4$, we show there is at most one solution of the problem and we give a quasi-reversibility method which converges uniformly on compact subsets of $(0,T]$ if and only if there exists a solution.

The plan is as follows:

I is a discussion of the relation between solutions of (E) and the semigroup generated by $-A$.

II introduces the QR-semigroups which describe our quasi-reversibility method.

III contains applications to certain parabolic evolution equations of second order in time.

I. THE SEMIGROUP AND SOLUTIONS

We shall assume that the linear operator A is maximal accretive and D(A) is dense in H. This is equivalent to each of the following [4,5,10]:

(a) $\mathrm{Re}(Ax,x) \geq 0$, $x \in D(A)$, and $I + A$ is onto H;

(b) $J_\alpha = (I + \alpha A)^{-1}$ is a contraction (defined everywhere) on H for each $\alpha > 0$;

(c) $-A$ generates a strongly-continuous semigroup $\{S(t): t \geq 0\}$ of contractions on H:

(i) $S(\cdot)x$ is continuous for each $x \in H$,

(ii) $S(t+s)=S(t)S(s)$, $S(0)=I$,

(iii) $\|S(t)\|_{\mathcal{L}(H)} \leq 1$,

(iv) $D(A) = \{x: \lim_{h\to 0} h^{-1}(S(h)x-x)$ exists$\}$, and the limit in H is $-Ax$.

Definition. A solution of (E) on $[a,b]$ is a function $u \in C([a,b],H) \cap C^1((a,b),H)$ for which $u(t) \in D(A)$ and (E) holds for all $t \in (a,b)$. It follows that u is a solution of (E) on $[a,b]$ if and only if $u(t)=S(t-a)u(a)$, $a \leq t \leq b$, and $u(t) \in D(A)$, $a < t < b$.

Definition. A weak solution of (E) on $[a,b]$ is a function of the form $u(t)=S(t-a)\xi$ for some $\xi \in H$.
Thus, the semigroup S generated by $-A$ is precisely the operational representation of (weak) solutions of (E) in terms of initial values.

Remark 1. There exists a weak solution of the final value problem if and only if $f=S(T)\xi$ for some $\xi \in H$.

Eventually we shall restrict our attention to those operators A as above which also satisfy

$$\mathrm{Re}(Ax,x) \geq |\mathrm{Im}(Ax,x)|, \qquad x \in D(A).$$

Then A is m-sectorial with angle $\pi/4$ [5] and the semigroup $\{S(t)\}$ is holomorphic. This implies that $S(t)x$ is (infinitely) differentiable at each $t > 0$, so every weak solution is a solution.

Remark 2. If $S(\cdot)$ is holomorphic and if the final value problem is properly posed, then a is bounded [2]. Thus in "most" situations to which our results apply the final value problem is necessarily improperly posed.

We sketch Yosida's elegant proof of the generation theorem [10]. Define the bounded operators $A_\alpha = AJ_\alpha$, $\alpha > 0$, and note that $A_\alpha = \alpha^{-1}(I-J_\alpha)$. Since J_α is a contraction, $\|A_\alpha x\| \leq \|Ax\|$

and $\|J_\alpha x - x\| \le \alpha\|Ax\|$, $x \in D(A)$, and hence $\|A_\alpha x - Ax\| \le \alpha\|A^2x\|$, $x \in D(A^2)$. These show A_α approximates A and J_α approximates I for small α. Each A_α is accretive so the group $S_\alpha(t) \equiv \exp(-A_\alpha t)$ consists of contractions for $t \ge 0$. These facts are used to prove the existence of the strong limit $S(t) \equiv \lim_{a \to 0} S_\alpha(t)$, $t \ge 0$, thereby defining the semigroup $S(\cdot)$.

II. QUASIREVERSIBILITY

In hopes of obtaining an approximate solution of the final value problem, we first solve the properly posed problem

$$v_\alpha'(t) + \alpha A v_\alpha'(t) + A v_\alpha(t) = 0 \qquad (E^\alpha)$$

$$v_\alpha(T) = f$$

for small $\alpha > 0$. Then we use $v_\alpha(0)$ as the initial value to determine a solution u_α of (E) with $u_\alpha(0) = v_\alpha(0)$. We expect to have $u_\alpha(T)$ close to f for sufficiently small $\alpha > 0$.

Note that (E^α) is equivalent to (E) with A replaced by the bounded operator A_α. Thus, we have $v_\alpha(t) = S_\alpha(t-T)f$, a representation by the group S_α, and our approximate solution to the final value problem is given by

$$u_\alpha(t) = S(t)S_\alpha(-T)f, \qquad 0 \le t \le T.$$

Our goal above is to show that $S(T)S_\alpha(-T)f$ is close to f. This suggests a Definition. For $\alpha > 0$, let $E_\alpha(t) = S(t)S_\alpha(-t)$, $t \ge 0$. $\{E_\alpha(\cdot)\}$ is the collection of QR-semigroups for the operator A. The QR-semigroups are stable if they are all contractions.

Lemma 1. $E_\alpha(\cdot)$ is generated by $-(A-A_\alpha)$.

Lemma 2. The following are equivalent:

(a) $\{E_\alpha(\cdot)\}$ is stable;

(b) $A-A_\alpha$ is accretive for every $\alpha > 0$;

(c) A^2 is accretive;

(d) $\mathrm{Re}(Ax,x) \ge |\mathrm{Im}(Ax,x)|$, $x \in D(A)$

Consider $\lim_{\alpha\to 0} E_\alpha(t)x$ for $x \in D(A)$. The Fundamental Theorem Calculus gives

$$E_\alpha(t)x - E_\beta(t)x = \int_0^t \frac{d}{ds}\left\{ E_\alpha(s)E_\beta(t-s)x \right\} ds$$
$$= \int_0^t E_\alpha(s)E_\beta(t-s)\,(A_\beta x - A_\alpha x)\, ds,$$

so when $\{E_\alpha(\cdot)\}$ are stable we obtain

$$\| E_\alpha(t)x - E_\beta(t)x \| \le t \| A_\beta x - A_\alpha x \|, \quad x \in D(A).$$

Hence we can define $E(t)x$ as the limit of $E_\alpha(t)x$ for $\alpha\to 0$ for $x \in D(A)$ and extend by continuity to $x \in H$. The convergence is uniform on bounded intervals, so we can take the limit in the integral

$$\int_0^t E_\alpha(s)(Ax - A_\alpha x)ds = x - E_\alpha(x), \quad x \in D(A)$$

to obtain $E(t)x=x$, hence $E(t) = I$. The preceding remarks indicate a proof of our

<u>Theorem 1</u>. In the situation of Lemma 2, $E_\alpha(t)x \to x$ (strongly) as $\alpha \to 0$ for $x \in H$, $t > 0$; the convergence is uniform on bounded intervals, and

$$\| E_\alpha(t)x - x \| \le t \| Ax - A_\alpha x \|, \quad x \in D(A).$$

<u>Corollary 1</u> (Backward Uniqueness). There is at most one solution of the final value problem.

<u>Proof</u>. By Remark 1 and linearity, this is equivalent to showing that the kernel of $S(T)$ is $\{0\}$. This is equivalent to showing the range of the adjoint $S^*(T)$ is dense in H. But the adjoint QR-semigroups $\{E_\alpha^*(t)\}$ are stable exactly when $\{E_\alpha(t)\}$ are stable, so Theorem 1 shows $S^*(T)S_\alpha^*(-T)x \to x$, hence the range of $S^*(T)$ is dense, and we are done.

Suppose $f=S(\delta)\xi$. Theorem 1 shows $\xi = \lim_{\alpha\to 0} E_\alpha(\delta)\xi = \lim_{\alpha\to 0} S_\alpha(-\delta)f$. Conversely if $\xi = \lim_{\alpha\to 0} S_\alpha(-\delta)f$, then each $S_\alpha(\delta)$ being a contraction implies $\lim_{\alpha\to 0} S_\alpha(\delta)S_\alpha(-\delta)f = S(\delta)\xi$. But this limit is f by Theorem 1. These observations and Remark 1 give us

Corollary 2 (Existence). Let $0 \leq \delta \leq T$. There is a solution u of (E) on $[T-\delta, T]$ with $u(T) = f$ if and only if $\lim_{\alpha\to 0} S_\alpha(-\delta)f$ exists in H. (In that case, the limit is precisely $u(T-\delta)$.)

In the situation of Corollary 2, we have the representations

$$u(t) = S(t + \delta - T)\xi, \qquad t \geq T - \delta,$$

$$u_\alpha(t) = S(t + \delta - T)E_\alpha(t)\xi, \qquad t \geq T - \delta,\ \alpha > 0$$

and hence derivatives of the difference are given by

$$u_\alpha^{(m)}(t) - u^{(m)}(t) = (-A)^m S(t+\delta-T)(E_\alpha(T)\xi - \xi),$$

$$\alpha > 0,\ t > T-\delta,\ m \geq 0.$$

Since $S(\cdot)$ is holomorphic, we obtain

Corollary 3 (Estimates). If there is a solution u of the final value problem on $[T-\delta, T]$, then

$$\| u_\alpha^{(m)}(t) - u^{(m)}(t) \| \leq [M/(t+\delta-T)]^m \| E_\alpha(T)\xi - \xi \|,$$

$$\alpha > 0,\ t > T-\delta,\ m \geq 0$$

and

$$\| u_\alpha^{(m)}(t) - u^{(m)}(t) \| \leq [M/(t+\delta-T-\epsilon)]^m T\alpha \| A^2 S(\epsilon)\xi \|,$$

$$\alpha > 0,\ 0 < \epsilon \leq \delta,\ t > T-\delta+\epsilon.$$

Remark 3. The quasi-reversibility method was introduced by Lattes and Lions [6]. They approximated (E) by the equation

$$w'(t) + Aw(t) - \alpha A^2 w(t) = 0$$

where A is self-adjoint and positive. See [7] for additional results and references.

Remark 4. When A is a realization of an elliptic partial differential operator, (E^{α}) is a pseudo-parabolic or Sobolev partial differential equation [8]. Such equations arise in various applications in which α corresponds to viscosity. This writer and Ting [9] observed that Yosida's proof shows that such equations approximate the corresponding parabolic equation (E). We may regard this approximation as a method of vanishing viscosity.

Remark 5 By considering solutions which satisfy a prescribed global bound, one can use the logarithmic convexity of solutions of (E) to stabilize the final value problem [1].

III. A SECOND ORDER EVOLUTION EQUATION

We attempt to apply our preceding results to the equation

$$v''(t) + Cv'(t) + Bv(t) = 0$$

where (for simplicity) B is self-adjoint and accretive in a Hilbert space $\mathcal{H}$, and C is accretive. (If -C is accretive, the final value problem is properly posed.) The change of variable $w(t) = e^{-\lambda t} \cdot v(t)$ gives the equivalent equation

$$\frac{d}{dt}\begin{pmatrix} w \\ w' \end{pmatrix} + \begin{pmatrix} 0 & -1 \\ \lambda^2+\lambda C+B & 2\lambda+C \end{pmatrix}\begin{pmatrix} w \\ w' \end{pmatrix} = \begin{pmatrix} 0 \\ 0 \end{pmatrix} .$$

Setting $\bar{u}(t) = e^{-\mu t}\begin{pmatrix} w \\ w' \end{pmatrix}$ gives us the equation (E) on the product space $H = \mathcal{H} \times \mathcal{H}$ with the operator

$$A = \begin{pmatrix} \mu & -1 \\ \lambda^2+\lambda C+B & \mu+2\lambda+C \end{pmatrix}$$

whose square is given by

$$A^2 = \begin{pmatrix} \mu^2-\lambda^2-(\lambda C+B) & -(2\mu+2\lambda+C) \\ (\lambda^2+\lambda C+B)(2\mu+2\lambda+C) & (\mu+2\lambda+C)^2-(\lambda^2+\lambda C+B) \end{pmatrix}$$

The difficulty with the preceding formalities is that it may be impossible, in general, to choose μ and λ so as to make A and A^2 accretive. (Consider A^2 with $\mu = \lambda = 0$ to appreciate the difficulty). In any event, such matrix operators almost always lead to technical difficulties.

We consider the extremely special case C=B, and take comfort in the fortunate fact that such examples do occur, e.g. in hydrodynamics and visco-elasticity [3,6]. Setting $\lambda=-1$ and $\mu=3$ in the above gives

$$A = \begin{pmatrix} 3 & -1 \\ 1 & 1+B \end{pmatrix} \qquad A^2 = \begin{pmatrix} 8 & -(4+B) \\ 4+B & B(2+B) \end{pmatrix}$$

Then, A satisfies the hypotheses of Theorem 1 and we have our final result.

<u>Theorem 2</u>. Let B be self-adjoint and accretive on the Hilbert space $\mathcal{H}$, and $f, g \in \mathcal{H}$. There is at most one solution $v \in C^1([0,T],\mathcal{H}) \cap C^2((0,T),\mathcal{H})$ of

$$v''(t) + Bv'(t) + Bv(t) = 0,$$

with $v'(t) + v(t) \in D(B)$ for $0 < t < T$ and

$$v(T) = f, \; v'(T) = g.$$

This problem is equivalent to the final value problem for (E) with A given above, $\bar{u}(t) = e^{-2t}\begin{pmatrix} v \\ v'+v \end{pmatrix}$, and $\bar{f} = e^{-2T}\begin{pmatrix} f \\ f+g \end{pmatrix}$.

Thus, for $0 \leq \delta \leq T$, there exists a solution v as above on the interval $[T-\delta,T]$ if and only if $\lim_{\alpha\to 0} \bar{u}_\alpha(-\delta)$ exists in $\mathcal{H}\times\mathcal{H}$ where $\bar{u}_\alpha$ is the solution of the approximating system

$$(1+3\alpha)u_1' + 3u_1 - \alpha u_2' - u_2 = 0,$$

$$\alpha u_1' + u_1 + [1+\alpha(1+B)]u_2' + (1+B)u_2 = 0,$$

$$u_1(T) = e^{-2T}f, \; u_2(T) = e^{-2T}(f+g).$$

<u>Remark 6</u> Similar results should hold in more general situations. The above proof technique might extend to the case where C

dominates B.

Remark 7. The above procedure approximates the equation by one of the form

$$B_1(\alpha)v''(t) + B_2(\alpha)v'(t) + B_3(\alpha)v(t) = 0$$

in which each $B_j(\alpha)$ is a polynomial in α and B of second degree in α and first degree in B. A simpler approximation is the Sobolev regularization.

$$(1+\alpha B)v''(t) + Cv'(t) + Bv(t) = 0$$

where we can assume without loss of generality that B dominates C. Such examples appear in fluid mechanics where $B = -\Delta$ and $\alpha > 0$ corresponds to inertia. (C.f., Remark 4.)

Remark 8. The techniques of this section require that we not make self-adjointness assumptions in Theorem 1: our matrix operator A is never self-adjoint.

REFERENCES

1. R. Ewing, The approximation of certain parabolic equations backward in time by Sobolev equations.

2. J. Goldstein, Some remarks on infinitesimal generators of analytic semigroups, Proc. Amer. Math. Soc., 22(1969), 91-93

3. J. Greenberg, R. MacCamey and V. Mizel, On the existence, uniqueness, and stability of solutions of the equation $\sigma'(u_x)u_{xx} + \lambda u_{xtx} = \rho.u_{tt}$, J. Math.Mech., 17 (1968), 707-728.

4. E. Hille and R. Phillips, Functional Analysis and Semi-groups, Colloquium Publications, vol. 31, Amer. Math. Soc. New York, 1957.

5. T. Kato, Perturbation Theory for Linear Operators, Springer-Verlag, Berlin-Heidelberg-New York, 1966.

6. R. Lattes et J.-L. Lions, Methode de Quasi-Reversibility et Applications, Dunod, Paris, 1967; English transl., American Elsevier, 1969.

7. L. Payne, Some general remarks on improperly posed problems for partial differential equations, Symposium on Non-Well-Posed Problems and Logarithmic Convexity, Lecture Notes in Mathematics, 316, Springer-Verlag, Berlin-Heidelberg-New York, 1973, 1-30.

8. R. Showalter, Existence and representation theorems for a semilinear Sobolev equation in Banach Space, SIAM J. Math. Anal., 3(1972), 527-543.

9. R. Showalter and T.W. Ting, Pseudo-parabolic parabolic partial differential equations, SIAM J. Math. Anal.,1 (1970), 1-26.

10. K. Yosida , Functional Analysis, Springer-Verlag, Berlin-Heidelberg-New York, 1965.

Research supported in part by National Science Foundation Grant GP-34261.

Department of Mathematics,
The University of Texas,
Austin, Texas 78712.

J R CANNON

A Class of Inverse Problems: The Determination of Second Order Elliptic Partial Differential Operators from Over-Specified Boundary Data

0. INTRODUCTION

Let $x = (x_1,...,x_n)$ denote a point in n-dimensional Euclidean space E_n and let Ω denote a domain contained in E_n with boundary $\partial\Omega$. Let $\mathcal{E}$ denote the class of functions $F=F(x,q,r,S)$, where $r \in E_n$, $S=(S_{ij}) \in E_{n^2}$, F is defined over $\Omega \times E_1 \times E_n \times E_{n^2}$, the partial derivatives $\frac{\partial F}{\partial S_{ij}}$ exist, and the matrix$(\frac{\partial F}{\partial S_{ij}})$ is positive definite. A reasonably general form of the inverse problem announced in the title can be stated as follows:

Problem: *Find a function* $F \in \mathcal{E}$ *and a function* $u=u(x)$ *defined over* Ω *such that*

(a) $F(x,u(x), \nabla u, (\frac{\partial^2 u}{\partial x_i \partial x_j}(x)))=0, \quad x \in \Omega,$

(b) $u(x)=f(x), \quad x \in \partial\Omega,$

(c) $G(x,f(x), \nabla u(x))=0, \quad x \in \partial\Omega,$

where ∇ *denotes the gradient operator in* E_n, f *is a known function defined over* $\partial\Omega$, and G *is a known function defined over* $\partial\Omega \times E_1 \times E_n$. In most applications G represents a statement concerning the flux and is known along with the form of F via some "law" such as Fourier's law, Fick's law and Darcy's law.

The purpose of this paper is to give the reader some insight into the difficulties of the problem and any specific simplification through a finite sequence of examples which arise naturally in physical applications. As shown below, both well-posed and not well-posed problems can arise in specific simplifications. The titles of the sections that follow refer to the physical application generating the specific simplification of the problem. Each section is self-contained and independent of any other section. The paper is concluded with a discussion (not a solution) of the problem and open questions of interest.

1. DETERMINATION OF SOURCES AND SINKS: GENERAL CASE

For this discussion it suffices to consider the subclass of functions of $\mathcal{E}$ which are of the form $F \equiv \sum_{i=1}^{n} S_{ii} - h(x)$ and the function $G = \frac{\partial u}{\partial n}(x) - g(x)$. For these choices, the problem simplifies to the determination of $u=u(x)$ and $h=h(x)$ such that

$$\begin{aligned} \Delta u &= h \quad , \quad x \in \Omega , \\ u &= f \quad , \quad x \in \partial\Omega, \\ \frac{\partial u}{\partial n} &= g \quad , \quad x \in \partial\Omega, \end{aligned} \tag{1.1}$$

where Δ denotes the Laplacian in E_n and $\frac{\partial}{\partial n}$ denotes differentiation in the direction of the outer normal. Since the problem is linear in u and h, it is easy to see that the uniqueness problem reduces to $f \equiv g \equiv 0$ implying that $u \equiv h \equiv 0$. However, it is clear that if u is any smooth function with compact support in Ω and if h is defined to be Δu, then (1.1) with $f \equiv g \equiv 0$ is satisfied by a multitude of u and h which are not identically zero. Consequently, to obtain a well-posed problem, additional restrictions upon the form of F are necessary.

2. DETERMINATION OF SOURCES AND SINKS: A RESTRICTED CASE

For the discussion here, let x_n be denoted by y and let $(x,y)=(x_1,\ldots,x_{n-1}, y)$ denote a point in E_n. Let Ω_1 denote a bounded domain in E_{n-1} and let $\Omega=\Omega_1 \times \{y: 0 < y < \infty\}$. Also, let D denote a domain (relative topology) which is contained in $\partial\Omega_1$. Consider now the problem of determining $u=u(x,y)$ and $h=h(x)$ (no dependence on y) such that

$$\sum_{i=1}^{n-1} \frac{\partial^2 u}{\partial x_i^2} + \frac{\partial^2 u}{\partial y^2} = h(x) \ , \ (x,y) \in \Omega \ , \tag{2.1}$$

$$u(x,y) = f(x,y), \quad (x,y) \in \partial\Omega,$$

$$\frac{\partial u}{\partial n}(x,y) = g(x,y), \quad (x,y) \in D \times (a,b), \tag{2.1}$$

where $0 < a < b < \infty$.

In order to demonstrate uniqueness here, let (u_i, h_i) $i = 1,2$ denote two solutions and set $w=u_1-u_2$ and $h^*=h_1-h_2$. Then, w and h^* satisfy

$$\sum_{i=1}^{n-1} \frac{\partial^2 w}{\partial x_i^2} + \frac{\partial^2 w}{\partial y^2} = h^*(x), \quad (x,y), \in \Omega,$$

$$w = 0, \quad (x,y) \in \partial\Omega, \tag{2.2}$$

$$\frac{\partial w}{\partial n} = 0, \quad (x,y) \in D \times (a,b).$$

Clearly,

$$w = \sum_{k=1}^{\infty} (-1) c_k (\lambda_k^{-1} - \lambda_k^{-1} e^{-\lambda_k^{\frac{1}{2}} y}) \psi_k, \tag{2.3}$$

where λ_k and ψ_k, $k=1,2,\ldots$, are the eigenvalues and normalized eigenfunctions, respectively, for

$$\Delta\psi + \lambda\psi = 0 \text{ in } \Omega_1,$$

$$\psi = 0 \text{ on } \partial\Omega_1, \tag{2.4}$$

and where

$$c_k = \int_{\Omega_1} \psi_k h^* dx. \tag{2.5}$$

From (2.2) and (2.3), it follows that for every x in D,

$$0 = \sum_{k=1}^{\infty} (-1) c_k (\lambda_k^{-1} - \lambda_k^{-1} e^{-\lambda_k^{\frac{1}{2}} y}) \frac{\partial \psi_k}{\partial n}. \tag{2.6}$$

At each x in D, $\frac{\partial w}{\partial n}$ is an analytic function of the form

$$\alpha_o + \sum_{k=1}^{\infty} \alpha_k e^{-\lambda_k^{\frac{1}{2}} y} .$$

From (2.6), it follows that

$$c_k \lambda_k^{-1} \frac{\psi_k}{\partial n} = 0, \; k = 1,2,\ldots . \tag{2.7}$$

Since $\frac{\partial \psi_k}{\partial n}$ cannot be identically zero in D it follows that

$$c_k = 0, \quad k = 1,2,\ldots$$

whence follows $h_1 \equiv h_2$ and $u_1 \equiv u_2$.

To gain further insight into problem (2.1), we consider n = 2 and the problems

$$\begin{aligned}
&\frac{\partial^2 z}{\partial x^2} + \frac{\partial^2 z}{\partial y^2} = h(x), \; 0 < x < 1, \; 0 < y < \infty , \\
&z(0,y) = z(1,y) = 0 \qquad , \; 0 < y < \infty , \\
&\frac{\partial z}{\partial x}(0,y) = (-1) \frac{a_n}{n\pi}(1-e^{-n\pi y}), \; \tfrac{1}{2} < y < 1, \\
&z(x,0) = 0 \qquad , \; 0 \leq x \leq 1.
\end{aligned} \tag{2.8}$$

For each positive n, the solution (z_n, h_n) is

$$h_n(x) = a_n \sin n\pi x \tag{2.9}$$

and

$$z_n(x,y) = (-1) \cdot \frac{a_n}{n^2 \pi^2} (1-e^{-n\pi y}) \sin n\pi x. \tag{2.10}$$

Choosing $a_n = \sqrt{n}$, a sequence h_n is obtained whose sup and L_2 norms tend to infinity while the data $\frac{\partial z_n}{\partial n}$ (0,y) tends uniformly to zero. Consequently, with respect to the norms usually considered in practical applications, problem (2.1) is not well-posed in the sense of Hadamard.

For additional information as to methods of numerically approximating solutions of problems analogous to (2.2) the reader is referred to [1].

3. DETERMINATION OF A CONDUCTIVITY $A(x)$

For problems of this nature, the subclass of function of $\mathcal{E}$ of the form $F \equiv A(x) \sum_{i=1}^{n} S_{ii} + \sum_{i=1}^{n} r_i \frac{\partial A}{\partial x_i}(x)$ are considered along with $G \equiv A(x)\frac{\partial u}{\partial n} - g(x)$. This yields the problem of determinating $u=u(x)$ and $A=A(x)$ satisfying

$$\begin{aligned} &\nabla \cdot A(x)\nabla u = 0, \quad x \in \Omega, \\ &u = f(x), \quad x \in \partial\Omega, \\ &A(x)\frac{\partial u}{\partial n} = g(x), \quad x \in \partial\Omega. \end{aligned} \tag{3.1}$$

For the special case of $n = 1$ and Ω the interval (a,b), the problem (3.1) becomes

$$\begin{aligned} &(A(x)u'(x))' = 0 \text{ in } (a,b), \\ &u(a) = \alpha, \quad u(b) = \beta, \\ &-A(a)u'(a) = -\gamma, \quad A(b)u'(b) = \gamma. \end{aligned} \tag{3.2}$$

where α, β, and γ are constants. Hence it follows that for any positive continuously differentiable function $B(x)$ which is defined over (a,b),

$$A(x) = \frac{\gamma \int_a^b B(\xi)d\xi}{(\beta-\alpha)B(x)} \tag{3.3}$$

and

$$u(x) = \alpha + (\beta-\alpha)\,\frac{\int_a^x B(\xi)\,d\xi}{\int_a^b B(\xi)\,d\xi}$$

form an acceptable solution of (3.2) provided $\gamma > 0$ when $\beta > \alpha$ and $\gamma < 0$ when $\beta < \alpha$. Thus, unique solvability for (3.1) cannot be expected in general.

Restricting A and u to have parallel gradients leads to a unique solvability theorem involving the level surfaces of certain harmonic functions. As this restriction is not physically interesting, the reader is referred to [2] for further details.

4. DETERMINATION OF A CONDUCTIVITY A(x): A RESTRICTED CASE

Consider the unit square in the Euclidean x-y plane and the problem of determining a bounded function $u=u(x,y)$ and a positive function $A=A(y)$ such that

$$\begin{aligned}
&(a)\quad \nabla\cdot A(y)\nabla u = 0,\ 0<x<1\ ,\ 0<y<1\ ,\\
&(b)\quad u(0,y) = f_1(y)\ ,\ 0<y<1\ ,\\
&(c)\quad u(1,y) = f_2(y)\ ,\ 0<y<1\ ,\\
&(d)\quad A(0)\frac{\partial u}{\partial y}(x,0) = g_1(x)\ ,\ 0<x<1\ ,\\
&(e)\quad A(1)\frac{\partial u}{\partial y}(x,1) = g_2(x)\ ,\ 0<x<1\ ,\ \text{and}\\
&(f)\quad A(y)\frac{\partial u}{\partial x}(0,y) = h(y)\ ,\ 0<y<1\ ,
\end{aligned}\tag{4.1}$$

where f_1, f_2, g_1, g_2 and h are known functions of their respective arguments. In the special case that $f_1\equiv 0$, $f_2\equiv 1$, $g_1\equiv g_2\equiv 0$, and h is positive valued, a rather simple uniquely determined solution can be found. Namely,

$$u = x \text{ and } A(y) = h(y). \tag{4.2}$$

The proof is quite simple. By the maximum principle, $u = x$ is the only solution possible for any conceivable $A(y)$.

Consequently, A(y) is uniquely determined from (4.1)-(f). It is not difficult to see that for small g_1 and g_2, the functions u and A can be obtained from a fixed point of the mapping.

$$TA(y) = h(y)\{\frac{\partial u}{\partial x}(0,y;A)\}^{-1} , \tag{4.3}$$

where the dependence of u upon A has been displayed. An analysis of this kind has been done and the reader is referred to [3] for the details. It should be noted that here the solution depends continuously upon the data, and therefore, the problem (4.1) is well-posed.

5. DETERMINATION OF A CONDUCTIVITY A(u).

For problems of this nature, the subclass of functions of $\mathcal{E}$ of the form $F \equiv A(q) \sum_{i=1}^{n} S_{ii} + A'(q) \sum_{i=1}^{n} (r_i)^2$ are considered along with $G \equiv A(u(x))\frac{\partial u}{\partial n} - q(x)$. This reduces to the problem of determining u = u(x) and A=A(u) satisfying

$$\nabla \cdot A(u)\nabla u = 0 \text{ in } \Omega ,$$

$$u = f \text{ on } \partial\Omega \tag{5.1}$$

$$A(f)\frac{\partial u}{\partial n} = g \text{ on } \partial\Omega .$$

The solution is obtained simply by considering [4]

$$z = \int_0^u A(\xi)d\xi \tag{5.2}$$

whence it follows that

$$\Delta z = 0 \text{ in } \Omega ,$$

$$\frac{\partial z}{\partial n} = g \text{ on } \partial\Omega,$$

and that

$$A(f(s)) = z'(s)\{f'(s)\}^{-1} , \tag{5.4}$$

where s is arc length along some curve on $\partial\Omega$ connecting the points of maximum and minimum values of f. Once A is determined, u follows from (5.2). Clearly, this problem depends continuously upon the data. Hence, it is well-posed.

6. REMARKS AND OPEN QUESTIONS

Clearly, the general question posed in the introduction is far too broad. The examples show that all types of mathematical problems arise from simplication of it. In some sense it can be conjectured that well-posed problems arise when the data is taken parallel to the quantity that is to be determined. The reader should reflect upon sections 4 and 5 for the meaning of parallel. Also, it can be conjectured that not well-posed problems arise when the data is taken orthogonal to the quantity that is to be determined. Sections 1, 2 and 3 contain the essence of what is meant by orthogonal in this discussion.

With respect to open questions of interest, the equations

$$\Delta u + \emptyset(u) = 0$$

and

$$\Delta u + c(x)u = 0$$

occur frequently in physical applications and it would be of interest to determine an unknown $\emptyset$ and c from overspecified boundary data.

Some comments should be made relative to a vast amount of literature on parameter determination which exists in the engineering journals. For the most part, the papers concern themselves with methods for computing the quantity in question. These methods usually involve numerical algorithms for the partial differential equations and an optimization algorithm for obtaining "the" quantity which produces the "best" fit of the additional data. It should be noted that the presentation of results of extensive numerical experimentations can be quite useful and informative.

REFERENCES

1. J.R. Cannon, Determination of an unknown heat source from overspecified boundary data, SIAM J. Numerical Anal. Vol. 5 (1968), pp. 275-286.

2. J.R. Cannon, and J.H. Halton, The irrotational solution of an elliptic differential equation with an unknown coefficient, Proc. Camb. Phil. Soc., (1963), Vol. 59, pp. 680-682.

3. J.R. Cannon and Paul Duchateau, Determination of the conductivity of an isotropic medium, To appear in J. Math. Anal. and Appl.

4. J.R. Cannon, Determination of the unknown coefficient k(u) in the equation $\nabla \cdot K(u) \nabla u = 0$ from overspecified boundary data, J. Math. Anal. and Appl., Vol. 18 (1967), pp. 112-114.

This research was supported in part by the National Science Foundation and Texas Tech University.

The University of Texas
at Austin, 78712.

H A LEVINE

Nonexistence of Global Weak Solutions to Non-linear Wave Equations

In this work, we will present, mostly without proof, some nonexistence theorems for some classes of parabolic and wave equations. Since the proofs of the general results in the first part of the talk are easy modifications of the concavity arguments described in Professor Payne's lectures and in the literature [3, 4, 5, 7, 8], we shall omit them. The details of the arguments in the second part of the talk have been given by Kaplan [1] and Glassey [2] for strong (classical) solutions and by Levine [6] for weak solutions.

1. We consider, in this section, the following two problems. Here $\Omega \subset R^n$ is a bounded domain with a piecewise smooth boundary.

Problem I.

$$\frac{\partial^2 u}{\partial t^2} = \sum_{i=1}^{n} \frac{\partial}{\partial x_i} \left(|\text{grad } u|^{p-2} \frac{\partial u}{\partial x_i} \right) + u^{q+1} \quad p \geq 2, \ q > 0$$

$$u(x,0) = u_0(x), \ u_t(x,0) = v_0(x), \ x \in \Omega$$

$$u(x,t) = 0, \ (x,t) \in \partial\Omega \times [0,T)$$

and

Problem II.

$$\frac{\partial u}{\partial t} = \sum_{i=1}^{n} \frac{\partial}{\partial x_i} \left(|\text{grad } u|^{p-2} \frac{\partial u}{\partial x_i} \right) + u^{q+1} \quad p \geq 2, \ q > 0$$

$$u(x,0) = u_0(x) \ x \in \Omega$$

$$u(x,t) = 0 \ (x,t) \in \partial\Omega \times [0,T).$$

In both of these problems, the existence interval, $[0,T)$, may or may not be finite. We note, in passing, that Tsutsumi [9] studied problem II. Our nonexistence results concerning this problem are similar to his, and, in so far as we consider weak solutions and the abstract Problem II-A below, extend his results. Unlike him, however, we do not consider existence of global solutions.

Corresponding to the preceding problems, we can consider, at least formally, two abstract problems. In these problems $u: [0,T) \to H$, a real Hilbert Space while P, A*, A are linear operators which need not be bounded. Without going into a great deal of technical detail, we list their relevant formal properties.

i) $(x,Px) > 0$ for all $x \in D_P$, $x \neq 0$

ii) $(x,A^*Ay) = -B(Ax, Ay)$

where $B(X,Y)$ is a positive symmetric bilinear form defined on the range of A. We let $\mathcal{F}$, $\mathcal{G}$ be nonlinear functionals (potentials) defined respectively on the range and domain of A. We let F and G denote their Frechet derivatives and assume the existence of a constant $\beta > 0$ such that

iii) $(\beta+2)\mathcal{F}(Ax) - B(F(Ax),Ax) \geq (\beta+2)\mathcal{G}(x) - (G(x),x)$

For all $x \in H$ for which the quantities in this inequality are meaningful.

We consider the abstract problems

<u>Problem I-A</u>

$$P\frac{d^2u}{dt^2} = A^*[F(Au(t))] + G(u(t))$$

$$u(0) = u_0,\ u_t(0) = v_0 \text{ prescribed}$$

and

Problem II-A

$$P\frac{du}{dt} = A^*[F(Au(t))] + G(u(t))$$

$u(0) = u_0$ prescribed.

For our examples, of course, $Af = \text{grad } f$ and $A^*Af = \text{div grad } f = \Delta f$, while

$$\mathcal{F}(Af) = \frac{1}{p}\int_\Omega |\text{grad } f|^p dx,$$

$$\mathcal{G}(f) = \frac{1}{q+2}\int_\Omega (f(x))^{q+2} dx,$$

where $f \in H_0^1(\Omega)$. Here, for vector functions $\vec{f} = (f_1, f_2, \ldots, f_n)$ and $\vec{g} = (g_1, g_2, \ldots, g_n)$ with components $f_i, g_i \in H_0^1(\Omega)$,

$$B(\vec{f},\vec{g}) \equiv \int_\Omega \left(\sum_{i=1}^n f_i(x) g_i(x)\right) dx.$$

Precise details are given in [3] among other places. If we write out (iii) for this case we find that

$$(\beta+2-p)(q+2)\int_\Omega |\nabla f(x)|^p dx \geq (\beta-q)p \int_\Omega (f(x))^{q+2} dx$$

which is satisfied for all $f \in H_0^1(\Omega)$ if $p < q+2$ and $\beta = q$.

Since Problems I, II may not have classical solutions, we shall be forced to treat weak solutions. Formally, a weak solution to Problem I-A is one which satisfies the initial conditions, and, for all $\phi : [0,t) \to H$, sufficiently regular,

$$(P^{1/2}\phi(t),\ P^{1/2}u_t(t)) = (P^{1/2}\phi(0),\ P^{1/2}v_0)$$

$$+ \int_0^t [(P^{1/2}\phi_\eta, P^{1/2}\phi_\eta) + (\phi,\ G(u)) - B(A\phi,\ F(Au))]d\eta$$

where $\phi = u$ is admissable, as well as the weak "conservation" law

$$E_1(t) \equiv \frac{1}{2}(P^{1/2}u_t,\ P^{1/2}u_t) + \mathcal{F}(Au(t)) - \mathcal{G}(u(t))$$

$$\leq E_1(0)$$

$$\equiv \frac{1}{2}(P^{1/2}v_0, P^{1/2}v_0) + \mathcal{F}(Au_0) - \mathcal{G}(u_0).$$

A weak solution to Problem II-A is one which satisfies the initial condition, and for all sufficiently regular $\phi : [0,T) \to H$,

$$(P^{1/2}\phi, P^{1/2}u_t) = (P^{1/2}\phi_0, P^{1/2}u_0) + \int_0^t [(\phi, G(u)) - B(A\phi, F(Au)]d\eta$$

where $\phi = u$ is admissable as well as the "conservation" law

$$E_2(t) \equiv \int_0^t \|P^{1/2}u_\eta(\eta)\|^2 d\eta + \mathcal{F}(Au(t)) - \mathcal{G}(u(t))$$

$$\leq \mathcal{F}(Au_0) - \mathcal{G}(u_0) \equiv E_2(0).$$

We have the following theorem:

Theorem 1. Under the preceding hypotheses we have, for some T, $0 < T < \infty$,

i) $\liminf_{t\to T^-} (T-t)^{1/\alpha}(P^{1/2}u(t), P^{1/2}u(t)) > 0$

for weak solutions to Problem I-A if $E_1(0) < 0$ where $\alpha = \beta/4$.

ii) $\liminf_{t\to T^-} (T-t)^{1/\alpha} \int_0^t (P^{1/2}u(\eta), P^{1/2}u(\eta))d\eta > 0$

for weak solutions to Problem II-A if $E_2(0) < 0$ and $\alpha = \beta/2$.

For our examples, we find that $E_1(0) < 0$ if

$$\frac{1}{(q+2)} \int_\Omega u_0^{q+2}(x)dx > \frac{1}{p}\int_\Omega |\text{grad } u_0(x)|^p dx + \frac{1}{2}\int_\Omega v_0^2(x)dx$$

which, for any $v_0 \in H_0^1(\Omega)$ will be satisfied if $u_0 \in H_0^1(\Omega)$ is chosen with max $|u_0|$ sufficiently large ($p < q+2$). Clearly, for such u_0, $E_2^\Omega(0) < 0$ also.

Now since Ω is bounded, it follows for Problem I from

Hölder's inequality that for such u_0, v_0,

$$\liminf_{t \to T^-} (T-t)^{1/\alpha} \|u(\cdot,t)\|_p > 0$$

where $\| \ \|_p$ denotes the $\mathcal{L}^p$ norm, for $2 \le p \le \infty$, while for such choices of u_0, it follows for Problem II,

$$\liminf_{t \to T^-} (T-t)^{1/\alpha} [\max_{0 \le \eta \le t} \|u(\cdot,\eta)\|_p] > 0.$$

Each of these last two limits have as their consequence, the nonexistence of global solutions to Problems I, II.

The abstract results can be applied in several interesting cases.

i) The equations of nonlinear elasticity

ii) P a differential operator ($P = I-\Delta$)

iii) If $p = 2$, $A = -A^* = \Delta$ so $AA^* = -\Delta^2$

iv) G an integral operator, for example

$$G(f)(x) = f(x) \int_\Omega k(x,y) f^2(y) dy$$

so that

$$G(f) = \frac{1}{4} \iint_{\Omega \times \Omega} k(x,y) f^2(x) f^2(y) dy dx.$$

Several other examples have been given in [4,5].

2. In some cases, the concavity result cannot be applied, for in the examples, if $p = 2$, or $F(Au) = Au$, the positivity of $-AA^*$ is crucial in the concavity method. Moreover, in the case of indefinite $N = -AA^*$, the method of logarithmic convexity likewise fails if we have an equation of the form

$$u_{tt} + \phi(u_t) = \Delta u.$$

In some of these cases, however, we can obtain nonexistence results via an idea of Kaplan [2]. Kaplan used his idea to study the nonexistence of strong (classical) solutions to second order parabolic equations. Later Glassey [1] used Kaplan's idea to study strong solutions to nonlinear wave equations as well as to the preceding equation.

We have used the technique to study the nonexistence of weak solutions to various kinds of nonlinear equations in [6]. We shall content ourselves with a single example here and list some of the other results at the end of our paper.

We consider

Problem III

$$\frac{\partial^2 u}{\partial t^2} = \sum_{i,j=1}^{n} \frac{\partial}{\partial x_j}\left(a_{ij}(x)\frac{\partial u}{\partial x_i}\right) + f(u(x,t)) \text{ in } \Omega \times [0,T)$$

$$u(x,0) = u_0(x),\ u_t(x,0) = v_0(x);\ x \in \Omega$$

$$u(x,t) = 0;\ (x,t) \in \partial\Omega \times [0,T).$$

We assume

$$\mathcal{L}f = \sum_{i,j=1}^{n} (a_{ij}f_{,i})_{,j}$$

is uniformly elliptic in $\bar{\Omega}$. Let ψ, λ_1 denote the first eigenfunction and eigenvalue for

$$\mathcal{L}\psi + \lambda_1\psi = 0 \quad x \in \Omega$$

$$\psi(x) = 0 \quad x \in \partial\Omega.$$

The nodal line theorem of Courant assures us that we may assume that $\psi > 0$ on Ω and ψ may be normalized so that

$$\int_\Omega \psi dx = 1.$$

We shall assume that the point function $f(s)$ satisfies

i) $f(s)$ is convex on $-\infty < s < \infty$.

ii) Let $\mathcal{F}(s) = \int^{s} f(\eta)d\eta$ (so $f(s) = \mathcal{F}'(s)$).

There is s_1 such that

$$\mathcal{F}(s) - \frac{1}{2}\lambda_1 s^2$$

is nondecreasing on (s_1, ∞).

iii) For any $\epsilon > 0$,

$$\int_{s_1}^{\infty} [\mathcal{F}(s) - \frac{1}{2}\lambda_1 s^2 + \epsilon^2]^{-1/2} ds < \infty.$$

As our definition of a weak solution, we shall mean a function $u(x,t)$ on $\Omega \times [0,T)$ satisfying the initial and boundary conditions of Problem III such that for all C^2 functions $\phi(x,t)$ vanishing on $\partial\Omega \times [0,T)$, we have

$$\begin{aligned}\int_{\Omega} \phi(x,t)\frac{\partial u}{\partial t}(x,t)dx = & \int_{\Omega} \phi(x,0)v_0(x)dx \\ & + \int_0^t \int_{\Omega} \phi(x,\eta)f(u(x,\eta))\,dxd\eta \\ & + \int_0^t \int_{\Omega} [\frac{\partial}{-\partial\eta}(x,\eta)\ \frac{\partial u}{\partial\eta}(x,\eta) \\ & \qquad + u(x,\eta)\mathcal{L}\phi(x,\eta)]dxd\eta.\end{aligned}$$

<u>Here, unlike the results obtainable via concavity and convexity methods, it is not necessary to impose some type of "conservation" law</u>. Note that the definition does not require much in the way of differentiability of the solution.

We have

<u>Theorem 2</u>. <u>Let</u> u <u>be a weak solution to Problem</u> III <u>in the above weak sense</u>. Set

$$F(t) = \int_{\Omega} u(x,t)\psi(x)dx$$

and assume

$F(0) > s_1, \quad F'(0) > 0,$

<u>where</u>

$$F(0) = \int_\Omega u_0(x)\psi(x)dx, \quad F'(0) = \int_\Omega v_0(x)\psi(x)dx.$$

<u>Then, for some</u> T, $0 < T < \infty$ <u>and all</u> p, $1 \leq p \leq \infty$,

$$\lim_{t \to T^-} \left(\int_\Omega |u(x,t)|^p dx\right)^{1/p} = +\infty.$$

The proof hinges on the observation that

$$F'(t) = \int_\Omega \frac{\partial u}{\partial t}(x,t)\psi(x)dx$$

can be written as an integral of the form $\int_0^t \int_\Omega$ as we see if we put $\phi(x,t) = \psi(x)$ in the above definition of a weak solution. Thus,

$$F'(t) = F'(0) + \int_0^t\int_\Omega \psi(x)[f(u(x,\eta)) - \lambda_1 u(x,\eta)]dxd\eta.$$

From this and Jenson's inequality it follows that

$$F''(t) \geq -\lambda_1 F(t) + f(F(t)).$$

Therefore it follows as in the arguments of Glassey [1] and Levine [6] as given by Professor Payne thismorning that $F(t)$ is unbounded on a finite interval $[0,T)$. The remainder of the result follows from Hölder's inequality.

The method has much to recommend it, as well as some substantial drawbacks. We list some of both of these here.

1. The type of the equation is not as important as in the concavity method. For example, it applies to

$$u_{tt} = u_{xx} - u_{yy} + u^2,$$

on $[0,\pi] \times [0,\pi] \times [0,T)$. Here $\psi(x,y) = c \sin x \sin y$, where c is a normalizing constant.

2. We get $\mathcal{L}^p$ approach to infinity for $1 \leq p \leq \infty$. This leads us to the observation that we can study, in a Banach Space, ordinary differential equations of the form,

$$Pu_{tt} = Au + \tilde{f}(u)$$

where A,P map a subdomain of the Banach Space B into itself. We need an element x in B*, the dual of B, such that

$$A^*x + \lambda P^*x = 0$$

where A* and P* are dual to A, P and satisfies

$$\langle x, \tilde{f}(u(t))\rangle \geq f(\langle x, Pu(t)\rangle).$$

Here $\langle , \rangle$ denotes the pairing between B* and B and f is a real valued function satisfying the conditions of the preceding Theorem.

3. The class of nonlinearities is different.

4. The class of weak solutions is wider and no a priori assumption on the energy is needed.

5. In addition to being applicable to equations of the form $Pu_t = Au + \tilde{f}(u)$, it can be applied to equations of the form

$$u_{tt} = Au + \phi(u_t)$$

while the concavity - convexity methods cannot.

6. If we wish to prove the nonexistence of global <u>positive</u> solutions, then f need only be convex on $[0,\infty)$.

The basic drawback is of course the verification of the hypotheses in item 3 of the preceding list. For example, consider

$$u_{tt} = -\Delta^2 u + u^2.$$

The linear operator $A = -\Delta^2$ is not known to possess an eigenfunction on Ω of one sign for general regions Ω. In fact there is evidence to indicate that the first eigenfunction for the square vibrating plate may indeed change sign. This limits the application in practice to second order differential operators A. There is also little known about eigenvalues and eigenfunctions of one differential operator relative to another.

3. Professor Lax posed the following question during this lecture. His question was the following:
In Problem I, is the breakdown of the solution due to the crossing of the (solution dependent) characteristics (the first term on the right hand side of the equation) or to the second term? We can only answer this by appealing to Problem II where the characteristics are not solution dependent but are only planes parallel to the initial domain $\Omega \times \{0\}$ and where the breakdown of the solution is therefore due to the second term on the right hand side of the equation. This is thus most likely the case in Problem I as well.

REFERENCES

1. Glassey, R.T., Blow up theorems for nonlinear wave equations, Math. Z 132 (1973), pp 183-203.

2. Kaplan, S., On the growth of solutions to quasilinear parabolic equations, Comm. Pur. Appl. Math. 16 (1957), pp. 327-330.

3. Knops, R.J., Levine, H.A. and Payne, L.E., Nonexistence, instability and growth theorems for solutions to an abstract nonlinear equation with applications to elastodynamics. Arch. Rat. Mech. Anal. (in print).

4. Levine, H.A., Instability and nonexistence of global solutions to nonlinear wave equations of the form $Pu_{tt} = -Au + \mathcal{F}(u)$. Trans. Am. Math. Soc. 192 (1974) pp. 1-21

5. Levine, H.A., Some nonexistence and instability theorems for formally parabolic equations of the form $Pu_t = -Au + \mathcal{F}(u)$ Arch. Rat. Mech. Anal. 51 (1973), pp. 371-386.

6. Levine, H.A. On the nonexistence of global solutions to a nonlinear Euler-Poisson-Darboux equation, J. Math. Anal. Appl. (in print).

7. Levine, H.A., Nonexistence of global weak solutions to some properly and improperly posed problems of mathematical physics. Math. Annalen. (submitted).

8. Levine, H.A., and Payne, L.E., Some nonexistence theorems for initial-boundary value problems with non-linear boundary constraints. Proc. Am. Math. Soc. (in print).

9. Levine, H.A., and Payne, L.E., A nonexistence theorem for the heat equation with a nonlinear boundary condition and for the porous medium equation, backward in time. J.D.E. 16 (1974), pp 319-334.

10. Tsutsumi, M., Existence and nonexistence of global solutions to nonlinear parabolic equations. Res. Inst. Math. Sci. Kyoto, U. 8 (1972) pp. 211-229.

Department of Mathematics
University of Rhode Island
Kingston, Rhode Island, U.S.A.

M GHIL

The Initialization Problem in Numerical Weather Prediction

1. INTRODUCTION

Dynamical meteorology deals with the mathematical description of the motions of the atmosphere, their causes and effects. With the increase in storage and speed of electronic computers, more precise and detailed weather forecasts based on the equations of dynamical meteorology have become computationally feasible. Therewith, problems connected to numerical weather prediction (NWP) have come to play an expanding role in this field. The mathematical problems arising from this include among others the extension of the time span for reliable forecasts, production of local forecasts from global ones, and the initialization problem, to be described here.

Without entering into the details of the physical rationale for this, the set of equations most widely used at the present in dynamical meteorology for the global description of the large-scale circulation of the atmosphere are the so-called primitive equations. These are the Euler equations of fluid dynamics, where the vertical momentum equation is replaced, due to the shallowness of the atmosphere, by the hydrostatic assumption. To this an energy equation has to be added. The system then, in a form which would look most familiar to fluid dynamicists, is

$$\rho_t + u\rho_x + v\rho_y + w\rho_z + (u_x + v_y + w_z) = 0 \tag{1a}$$

$$u_t + uu_x + vu_y + wu_z = \frac{1}{\rho}p_x + fu \tag{1b}$$

$$v_t + uv_x + vv_y + wv_z = \frac{1}{\rho}p_y - fv \tag{1c}$$

$$0 = \frac{1}{\rho}p_z - g \tag{1d}$$

$$S_t + uS_x + vS_y + wS_z = 0 \qquad (1e)$$

$$p = p(\rho, S) \qquad (1f)$$

The equations above are written in cartesian coordinates x, y, z, rather than in spherical coordinates, in order to simplify the discussion and for the same reason no diffusion or source terms appear. However the Coriolis force (fu,-fv,0) where $f = 2\Omega \sin\phi$, Ω is the angular velocity of the earth, ϕ latitude, has to be included. Since the density ρ and the acceleration of gravity g are strictly positive, for fixed x,y equations (1d) and (1f) combined yield the pressure p as a strictly monotonic (decreasing) function of z

$$p_s - p = \int_0^z \rho(p(z), S(z))\, g(z)\, dz, \qquad (1f')$$

where $p_s(x,y)$ is the surface pressure, and the x,y dependence was suppressed throughout. Hence p can be taken as a new "vertical coordinate", and in fact g(z) is assumed to be constant. Moreover, it is customary instead of the entropy S to use the "potential temperature" θ, defined by

$$S = c_p \log \theta + \text{const.} \qquad (2a)$$

or, with a specific choice of constant,

$$\theta = \left(\frac{p_0}{p}\right)^{R/c_p} T \qquad (2b)$$

where T is temperature, R the gas constant, c_p specific heat at constant pressure and p_0 = 1000 mbar.

After these changes the basic system of equations, in "pressure coordinates" x,y,p,t, becomes

$$u_x + v_y + \omega_p = 0 \qquad (3a)$$

$$u_t + uu_x + vu_y + \omega u_p - fv = -\phi_x \qquad (3b)$$

$$v_t + uv_x + vv_y + \omega v_p + fu = -\phi_y \qquad (3c)$$

$$\theta_t + u\theta_x + v\theta_y + \omega\theta_p = 0 \tag{3d}$$

$$\phi_p = - RT/p, \tag{3e}$$

where the gas law

$$p/\rho = R\ T \tag{1f'}$$

was used to obtain (3e) from (1d), and where $\phi = g\ z$ is the geopotential of an isobaric surface, $\phi = \phi(x,y,p,t)$. Note that $\omega \equiv dp/dt$ formally replaces w in system (3). We shall again call (3a) - (3d) the primitive equations.

Although no existence and uniqueness theorems exist for system (1) or (3), it is reasonable to assume that a well posed problem for either system considered over the whole earth would be to prescribe the unknowns at time $t = t_0$, as well as certain boundary conditions at the upper and lower boundary of the atmosphere.

The nature of physically plausible boundary conditions is more or less understood. No major problems seem to arise from the numerical approximation of the boundary conditions presently used in integrations extending over a period of a few days. Though we shall touch upon this problem in Section 3, our subject will be the initial data for the primitive equations.

Specifically two problems arise, connected with (i) the completeness, and (ii) the accuracy of the initial data. In practice system (3), say, is solved numerically by finite-difference methods over a uniform grid. Such a finite difference formulation of (3) is the essential part of what is called a general circulation model (GCM).

Accuracy and scale considerations, together with computing capability limitations, make 2° - 5° a reasonable mesh size for a GCM. However at the present data gathering facilities are not distributed densely enough to provide initial data for a GCM at all grid points uniformly. In fact large"white patches", as well as many non-uniformities, exist in the meteorological data coverage, especially over oceans and in the southern hemisphere. This is the (lack of) completeness-of-data problem.

Moreover, when physically reasonable data are supplied at all grid points, as they have to be in any numerical experiment with a GCM, this does not seem to produce a physically credible solution. The problem is that the primitive equations have solutions which include both large scale, slowly moving phenomena (customarily associated with Rossby waves, these being exact solutions of a simpler, linearized system), and much faster, small amplitude phenomena (similarly associated with inertia-gravity waves) ([1]). In any numerical experiment, the truncation error of the scheme produces "inertia-gravity" waves of exaggerated amplitude, and in a straightforward integration these are reduced to reasonable size only after periods of a few days of simulated time. After comparable periods however the large scale solutions differs too much from the one obtained starting with slightly different initial data. This is the accuracy problem, in the sense that small errors in the initial data seem to produce inadmissibly large errors in the solution. Providing complete and sufficiently accurate initial data for a GCM constitutes what is called the initialization problem.

The idea in overcoming both problems relies on a fact which has been well known in meteorology for some time: namely, that certain time-independent relations, called balance equations, hold at least approximately for large scale atmospheric phenomena. This state of approximate "balance" is the result of the process of "geostrophic adjustment", in which inertia gravity waves are dispersed and dissipated ([4]). It was suggested that initial data which satisfied such balance equations would produce much smaller inertia-gravity waves.*

---

*This would be similar to a solution (E,B) of the Maxwell equations E_t - curl B = j, B_t + curl E = η which can be shown to satisfy at all times div B = 0, div E = η if these were satisfied initially.

Also it has been suggested that it is only necessary to supply some of the initial data which the system requires, since the balance equations should be used to obtain the rest, speaking vaguely for the moment.

It turned out however that this idea in the simple form outlined above did not work: "balanced" initial data still produced large inertia-gravity waves ([5] and [6]). The simple reason was that the balance equations used were not consistent with the primitive equations, since the former were based on certain approximations not made in deriving the latter.

The next step taken by dynamical meteorologists was to let the model take care of itself so to speak: provide the available, inaccurate data, and let the GCM simulate geostrophic adjustment. After a certain time of integration, the numerical solution would approach the correct one, and in particular it would become approximately balanced. This would happen provided of course that over the same time span partial data continue to be supplied, so as to "force" the adjustment.

In particular it appears that satellite technology holds promise for supplying in the future rather accurate temperature data over the whole earth. Given temperature and surface pressure, pressure can be obtained from equation (1f') or rather by integrating the ordinary differential equation

$$\frac{dp}{dz} = -g\rho(T(z),p), \quad p(0) = p_s, \tag{4}$$

with parameters x,y. Then we can similarly integrate for ϕ in (3e).

Now consider the following computer experiment: a "control" run consisting of a numerical integration of (3) is made with given initial data u_0, v_0, w_0 and T_0. Its results over a certain time span, u_c, v_c, w_c, T_c are considered for comparison purposes to be the "true history" of the atmosphere for that period. Then other "runs", or integrations, are made, in which the same initial temperature data $T(x,y,p,0) = T_0$, but slightly different initial velocity data $(u,v,w)|_{t=0}$ are used, and in which, instead of using $T(x,y,p,t)$ produced by the given run, one substitutes every few time steps the value of $T_c(t)$ from the control run (one"updates" the value of T). The procedure is considered successful if the results of a given run converge after a certain time to those of the control run ([2] and [8]).

The evidence is that convergence occurs, though rather slowly, for the procedure as described above, but divergence occurs when using for comparison a "control run" from a different GCM or real data ([5] and [6]). This and other disadvantages of the procedure led us to try to revert to the original idea of time-independent compatibility conditions, which should however be consistent with the primitive equations, i.e, should be derived from the latter. These compatibility conditions could include time-derivatives of the temperature T and of other thermodynamic fields (p,ρ,θ,ϕ), these fields being completely determined by T and the hydrostatic equation, but should not contain time derivatives of the field (u,v,w) or (u,v,ω). That is, the equations provided by these compatibility conditions would have to be solved for the velocity variables (u,v,w) at time $t = t_0$ ([3]).

We shall illustrate the idea first by a simple example, for which it can be shown analytically to be entirely successful.

2. INITIAL DATA FOR THE SHALLOW-FLUID EQUATIONS

Consider the shallow fluid equations for a rotating cartesian x,y-frame

$$u_t + uu_x + vu_y + \phi_x - fv = 0, \qquad (1a)$$

$$u_t + uv_x + vv_y + \phi_y + fu = 0, \qquad (1b)$$

$$\phi_t + u\phi_x + v\phi_y + \phi(u_x+v_y) = 0, \qquad (1c)$$

$$\phi = g\ h.$$

Here u, v are the velocity components in the x,y direction respectively, h is the height of the free surface and f, the coriolis parameter, is taken to be constant.

First we will study the linearization of this system around a state of rest

$$u_t + \phi_x - fv = 0, \tag{2a}$$

$$v_t + \phi_y + fu = 0, \tag{2b}$$

$$\phi_t + \Phi(u_x + v_y) = 0, \tag{2c}$$

$$\Phi = gH = \text{const.}$$

Here Φ is the equilibrium value of the geopotential gh of the free surface, and ϕ stands now for gh-Φ, the deviation of the geopotential from Φ.

The value of the study of (1) and even (2) as simpler models for the primitive equations consists in the fact that already (2) has three independent, physically significant plane wave solutions. One of these corresponds to slow Rossby waves, the other two to much faster inertia gravity waves propagating in opposite directions ([5], [7]). As we saw, phenomena with this character are recognized in the solutions of the primitive equations, and are even believed to be their most salient features. In the case at hand, where the unperturbed velocity is zero, the Rossby mode is stationary and the inertia-gravity waves have phase velocity

$$c = \pm (k^2\Phi + f^2)^{1/2}/k, \quad k^2 = k_1{}^2 + k_2{}^2,$$

(k_1, k_2) being the wave vector.* Any solution of (2) can be represented by a series expansion in these plane waves.

The pure initial value problem for (2) is well known to be properly posed. Our present viewpoint however is that initial data are available at time $t = t_0$ for ϕ and its time derivatives, but not for u, v.

*The terminology of inertia-gravity waves is used to indicate that $c^2 = gH + f^2/k^2$, the term f^2/k^2 resulting from the "inertial" coriolis force present in the equations, as opposed to "pure" gravity waves for which $c^2 = gH$.

The proposed approach is to try to derive from (2) compatibility conditions, in the form of equations for u,v, in which their time derivatives u_t, v_t would not appear and which therefore, together with adequate boundary conditions, would determine u, v uniquely at time $t = t_0$. Thus a complete set of initial data $(\phi, u, v)|_{t=t_0}$ would become available and determine the solution (ϕ, u, v) of (2) at future times.

This is easily done in the case of system (2). Differentiate (2c) with respect to t, (2a) and (2b) with respect to x and y respectively to yield

$$u_{tx} + \phi_{xx} - fu_x = 0 \tag{3a}$$

$$v_{ty} + \phi_{yy} + fu_y = 0 \tag{3b}$$

$$\phi_{tt} + \Phi(u_{xt}+v_{yt}) = 0. \tag{3c}$$

Substituting u_{xt} and v_{yt} from (3a), (3b) into (3c), we obtain an equation which together with (2c) forms the system*

$$u_x + v_y = -\phi_t/\Phi, \tag{4a}$$

$$u_y - v_x = (\phi_{tt}-\Phi\Delta\phi)/\Phi f. \tag{4b}$$

Here Δ is the two-dimensional Laplacian, $\Delta = \partial_x^2 + \partial_y^2$, and the right-hand sides are known by assumption. System (4) is elliptic and in fact leads by cross-differentiation to a Poisson equation for either u or v. A well-posed problem

---

*Notice that linearization around a solution with velocity (U,V) satisfying $fU = -\Phi_y$, $fV = \Phi_x$, $\Phi \neq$ const. would still allow us to eliminate the time derivatives of u, v by cross-differentiation in (2), but then second space derivatives of u, v would appear in the equation equivalent to (4b), and the analysis would become more complicated. A similar statement applies to letting f depend on y, say.

for (4) in a bounded domain D would be the Dirichlet problem i.e., prescribing u say on the boundary of D, ∂D, and v at some point in $\bar{D} = D \cup \partial D$.

This then solves the completeness of initial data problem for system (2). To analyse the accuracy problem, we have first to see what a balanced solution means in this instance. The meteorologically significant balanced state for this simple case is that of "geostrophic balance,"

$$fu = -\phi_y \ , \ fv = \phi_x. \tag{5}$$

Such a state is a solution of (2), as well as of (4), with $\partial_t \equiv 0$.

A solution of (5) corresponds to a Rossby-wave solution of (2). The most general stationary solution of (2) however will satisfy

$$u_x + v_y = 0, \tag{6a}$$

$$u_x = u_y = \Delta\chi, \tag{6b}$$

where $\chi = \phi/f$. The solution of (6a) is

$$u = -\psi_y \ , \ v = \psi_x, \tag{5'}$$

with ψ an arbitrary, twice continuously differentiable function. By (6b) ψ has to satisfy the equation

$$\Delta(\psi-\chi) = 0.$$

Hence if either

$$\psi = \chi \text{ on } \partial D$$

or

$$\partial_n\psi = \partial_n\chi \text{ on } \partial D,$$

with ∂_n the normal derivative, then

$$\psi = \chi + \text{const. in } D,$$

and (u,v) is uniquely determined by (5'), ϕ being known.

It easily follows that if the solution of (6) is "geostrophic" on ∂D, i.e., satisfies there (5), then it is so throughout D.

Assume now that certain small errors are made in measuring $\phi(t)$ in $\bar{D}$ and $(u,v)|_{t=t_0}$ on ∂D, or that for any other reason the measured state is not entirely geostrophic. As a consequence $\phi_t, \phi_{tt} \neq 0$ and ϕ will contain non-zero inertia-gravity waves in its plane wave expansion. We want however the solution $(\phi,u,v,)$ of (2) after initialization via (4) to be free from or contain only small-amplitude inertia gravity waves. To achieve this, set first the coefficients of the inertia-gravity waves in the expansion of ϕ to zero, and call the new function thus obtained ϕ'. Then $\phi'_t \equiv \phi'_{tt} \equiv 0$ and $(u,v)|_{t=t_0}$ will satisfy rather than (4), a system of the form (6), with $\chi^0 = \phi'/f$. On the boundary ∂D, however, the departure e from geostrophicity,

$$e(x,y) = \max \{|fu+\phi'|, |fu-\phi'_x|\}$$

will in general not be zero. But by the maximum principle

$$\max_{D} e \leq \max_{\partial D} e. \tag{7}$$

Thus it suffices to ensure that e satisfy on ∂D a certain prescribed smallness condition for that condition to hold throughout D. This settles also the accuracy question for system (2).

We turn now to the more general system (1) and ask for a set of compatibility conditions similar to (4) which would solve the completeness problem. In case such a set existed, one would then also want to derive an error bound for its solutions similar to (7). We will have, however, to stop short of this goal, since it will be seen that the set of conditions obtained is actually incomplete and does not determine

$(u,v)|_{t=t_0}$ uniquely, at least not under the present assumptions*.

To obtain compatibility conditions similar to (4) we have to eliminate u_t, v_t from (1), as we did in (2). First we simplify (1c) by introducing

$$\varphi = \log \phi.$$

Again it is assumed that ϕ, and hence φ, and their derivatives are known at $t = t_0$, but that u,v are not. Dividing through in (1c) by $\phi \geq \alpha > 0$ we obtain

$$u_x + v_y + \varphi_x u + \varphi_y v + \varphi_t = 0. \tag{8}$$

Differentiating this with respect to t and (1a), (1b) with respect to x,y we obtain

$$u_{tx} + uu_{xx} + vu_{xy} + u_x^2 + u_y v_x - fv_x + \phi_{xx} = 0 \tag{9a}$$

$$v_{ty} + uv_{xy} + vv_{yy} + u_y v_x + v_y^2 + fu_y + \phi_{yy} = 0 \tag{9b}$$

$$u_{xt} + v_{yt} + \varphi_x u_t + \varphi_y v_t + \varphi_{xt} u + \varphi_{yt} v + \varphi_{tt} = 0 \tag{9c}$$

Substituting into (9c) u_t, v_t from (1a), (1b), and u_{xt}, v_{yt} from (9a), (9b), yields an equation in u,v which contains no time differentiation of the latter,

$$\begin{aligned} &uu_{xx} + vu_{xy} + uv_{xy} + u_x^2 + 2u_y v_x + v_y^2 + f(u_y - v_x) \\ &+ \phi_x(uu_x + vu_y + \phi_x - fv) + \phi_y(uv_x + vv_y + \phi_y + fu) \\ &- \varphi_{xt} u - \varphi_{yt} v + \phi_{xx} + \phi_{yy} - \varphi_{tt} = 0 \end{aligned} \tag{10}$$

---

*System (1) is totally hyperbolic, having three distinct families of characteristic surfaces and the surface $t = t_0$ is free (non-characteristic) and clearly space-like. Hence the pure initial-value problem for (1) is well posed and complete and sufficiently accurate initial data (o,u,v) ought to produce a reasonable numerical solution.

Together with (8) this would then yield the desired set of compatibility conditions. Only one has to verify whether (8) and (10) are indeed independent equations. To do this we transform the system composed of (8) and (10) into a first order system. Let

$$u_y \equiv q, \; v_x \equiv r.$$

Then an equivalent first-order system is

$$
\begin{aligned}
&u_x = p \;, \\
&u_y = s \;, \\
&p_x + s_x + \varphi_x p + \varphi_y r + \varphi_{xx} u + \varphi_{xy} v = -\varphi_{tx}, \\
&p_y - q_x = 0 \;, \\
&s_x - r_y = 0 \;, \\
&up_x + vp_y + us_x + vs_y + p^2 + 2qr + r^2 + f(q-r) + \varphi_x (up+vq-fv) \\
&+ \varphi_y \; (ur+vs+fu) - \varphi_{xt} u - \varphi_{yt} v = \varphi_{tt} - \phi_{xx} - \phi_{yy} - \varphi_x \phi_x - \varphi_y \phi_y,
\end{aligned}
\qquad (11)
$$

where differentiation of (8) with respect to x led to the third equation.

Let a characteristic curve of (11) be given by

$$\psi(x,y) = \text{const.}$$

and define

$$\xi = \psi_x \;, \quad \eta = \psi_y, \quad \tau = u\xi + v\eta.$$

The characteristic equation of (11) is then

$$\det \begin{pmatrix} \xi & & & & & \\ & \eta & & & & \\ & & \xi & & & \xi \\ & & \eta & -\xi & & \\ & & & & -\eta & \\ & & & & & \tau \end{pmatrix} = 0 \quad . \tag{12}$$

But developing the determinant yields

$$\xi^2\eta(\xi\eta\tau - \xi\eta\tau) \equiv 0,$$

and hence (12) holds for any curve in the x,y-plane, ie., the equations of system (11) cannot be independent. Therefore (10) cannot be independent of (8).

The implications of this situation for the initialization problem will be discussed after carrying out the analogous analysis for the primitive equations, to which we now return.

3. INITIAL DATA FOR THE PRIMITIVE EQUATIONS

For the sake of simplicity we shall discuss the primitive equations in pressure coordinates x,y,p, i.e., equations (3a-d) of Section 1, which we write down again for convenience

$$u_x + v_y + \omega_p = 0 \tag{1a}$$

$$u_t + uu_x + vu_y + \omega u_p - fu = -\phi_x , \tag{1b}$$

$$v_t + uv_x + vv_y + \omega v_p + fu = -\phi_y , \tag{1c}$$

$$\theta_t + u\theta_x + v\theta_y + \omega\theta_p = 0. \tag{1d}$$

The same result can be shown to hold in x,y,z-coordinates, and they are also independent of the exact form of the known inhomogeneous terms in (1). As pointed out already in Section 1, knowing the temperature T and surface pressure p_s will allow us immediately to obtain the pressure p and all other thermodynamic variables via equations (4), (1f) and (3e) of Sec. 1.

Our assumption then is that, limiting our attention to system (1) above, θ, ϕ and all necessary derivatives are known at time $t = t_0$. We wish to derive from (1) compatibility conditions in which time derivatives of $W = (u,v,w)$ do not appear and investigate whether the set of these conditions will determine the wind field W at time $t = t_0$ uniquely*.

Since no equation containing ω_t exists in (1), it is natural to use (1d) to eliminate ω,

$$\omega = - \frac{1}{\theta_p} (\theta_x u + \theta_y v + \theta_t). \tag{2}$$

Differentiating this with respect to p and substituting into (1a) we obtain

$$u_x + \theta^{(1)} u_p + v_y + \theta^{(2)} v_p + \theta^{(3)} u + \theta^{(4)} v = -\theta^{(5)},$$

where

$$\theta^{(1)} = - \theta_x/\theta_p,$$

$$\theta^{(2)} = - \theta_y/\theta_p,$$

---

*Note that system (1) is hyperbolic, in the sense that it has four families of characteristic hypersurfaces (three of which coincide), but that the hyperplane $t = t_0$ is not free. Hence a Goursat problem, with part of the data prescribed on $t = t_0$, part on $p = p_0$, is probably adequate for (1), rather than a pure initial-value problem. In practice p_0 = const. cannot be taken to be zero and the conditions to be imposed on it are not self-evident either mathematically or physically. Therefore the following discussion is rather heuristic, assuming that the conditions actually used on $p = p_0$ in GCMs, together with θ, W at $t = t_0$ and the usual boundary conditions will ensure existence and uniqueness for integration times of a few days.

$$\theta^{(3)} = \theta_x \theta_{pp}/\theta_p{}^2 - \theta_{xp}/\theta_p,$$

$$\theta^{(4)} = \theta_y \theta_{pp}/\theta_p{}^2 - \theta_{yp}/\theta_p,$$

$$\theta^{(5)} = \theta_t \theta_{pp}/\theta_p{}^2 - \theta_{tp}/\theta_p$$

are known functions.

To eliminate the time derivatives of u, v between (3) and (1b), (1c) we differentiate (3) with respect to t and (1b), (1c) with respect to p and x, and with respect to p and y, to yield

$$u_{xt} + \theta^{(1)} u_{pt} + v_{yt} + \theta^{(2)} v_{pt} + \theta^{(3)} u_t + \theta^{(4)} v_t \qquad (3')$$

$$+ \theta_t{}^{(1)} u_p + \theta_t{}^{(2)} v_p + \theta_t{}^{(3)} u + \theta_t{}^{(4)} v = -\theta_t{}^{(5)},$$

$$u_{xt} + uu_{xx} + vu_{xy} + \omega u_{xp} + u_x^2 + u_y v_x + u_p \omega_x - f v_x = -\phi_{xx}, \qquad (4a)$$

$$u_{pt} + uu_{xp} + vu_{yp} + \omega u_{pp} + u_x u_p + u_y v_p + u_p \omega_p - f v_p = -\phi_{xp}, \qquad (4b)$$

$$v_{yt} + uv_{xy} + vv_{yy} + \omega v_{yp} + u_y v_x + v_y^2 + v_p \omega_y + f u_y = -\phi_{yy}, \qquad (4c)$$

$$v_{pt} + uv_{xp} + vv_{yp} + \omega v_{pp} + u_p v_x + v_y v_p + v_p \omega_p + f u_p = -\phi_{yp}, \qquad (4d)$$

Substituting u_{xt}, u_{pt}, v_{pt} from (4) and u_t, v_t from (1b), (1c) into (3') we obtain

$$uu_{xx} + vu_{xy} + \omega u_{xp} + uv_{xy} + vv_{yy} + \omega v_{yp} \qquad (5)$$

$$+ \theta^{(1)}(uu_{xp} + vu_{yp} + \omega u_{pp}) + \theta^{(2)}(uv_{xp} + vv_{yp} + \omega v_{pp})$$

$$+ u_x^2 + u_y v_x + u_p \omega_x + u_y v_x + v_y^2 + v_p \omega_y$$

$$+ \theta^{(1)}(u_x u_p + u_y v_p + u_p \omega_p) + \theta^{(2)}(u_p v_x + v_y v_p + v_p \omega_p)$$

$$+ \theta^{(3)}(uu_x + vu_y + \omega u_p) + \theta^{(4)}(uv_x + vv_y + \omega v_p) +$$

$$+ f(u_y - v_x) \pm f(\theta^{(2)}u_p - \theta^{(1)}v_p) - \theta^{(1)}u_p - \theta_t^{(2)}v_p$$
$$+ f(\theta^{(4)}u - \theta^{(3)}v) - \theta_t^{(3)}u - \theta_t^{(4)}v$$
$$= -\phi_{xx} - \phi_{yy} - \theta^{(1)}\phi_{xp} - \theta^{(2)}\phi_{yp} - \theta^{(3)}\phi_x - \theta^{(4)}\phi_y + \theta_t^{(5)}.$$

Here again ω and its first space derivatives can be eliminated by using (2) and only first order space derivatives of u,v appear in the process. Thus we can view (5) as a second order equation in u,v.

To test the independence of (3) and (5) we again convert to a first order system. Introducing

$$u_y = m,\ u_p = r,\ v_x = n,\ v_p = t,$$

and differentiating (3) with respect to p leads to the following equivalent system:

$$\begin{aligned}
&u_x = q,\\
&v_y = s,\\
&m_x - q_y = 0,\\
&n_y - s_x = 0,\\
&q_p - r_x = 0,\\
&s_p - t_y = 0,\\
&s_p - t_y = 0,\\
&uq_x + vq_y + (\omega + \theta^{(1)}u)q_p + \theta^{(1)}(vr_y + \omega r_p) + us_x + vs_y\\
&\quad + (\omega + \theta^{(2)}v)s_p + \theta^{(2)}(ut_x + \omega t_p) + \ldots = 0,\\
&r_x + \theta^{(1)}r_p + t_y + \theta^{(2)}t_p + \ldots = 0,
\end{aligned} \tag{6}$$

where the dots stand for terms which do not contain the derivatives of the unknowns $U = (u,v,m,n,q,r,s,t)$.

Let a surface in x,y,p-space be given by

$$\psi(x,y,p) = \text{const.}$$

and denote by

$$\xi = \psi_x , \quad \eta = \psi_y , \quad \zeta = \psi_p$$

the components of its normal. Reordering slightly the equations in (6) yields the characteristic equation

$$\det \begin{bmatrix} \xi & & & & & & & \\ & \eta & & & & & & \\ & & \xi & & -\eta & & & \\ & & & \eta & & & -\xi & \\ & & & & \zeta & -\xi & & \\ & & & & \tau+(\omega+\theta^{(1)}u)\zeta & \theta^{(1)}(v\eta+\omega\zeta) & \tau+(\omega+\theta^{(2)}v)\zeta & \theta^{(2)}(u\xi+\omega\zeta) \\ & & & & & \xi+\theta^{(1)}\zeta & & \eta+\theta^{(2)}\zeta \\ & & & & & & \zeta & -\eta \end{bmatrix} \tag{7}$$

$$\equiv \det A = 0$$

Developing the determinant in (7) again yields

$$\det A \equiv 0,$$

irrespective of x,y,p and the solution U on which A depends. Thus the equations of (6) cannot be all independent of each other and neither can equations (3) and (5).

The situation then is quite similar to that encountered in discussing the full, nonlinear shallow fluid equations in Sec. 2. There eq. (1c) was the only instantaneous condition at $t = t_0$ on the two unknowns u, v whereas here eqs. (1a), (1d) are the only such conditions on u, v, ω. Hence in both cases we seem to fall short by one equation.

The idea that temperature and surface pressure together with their "past history" completely determine the present state of the atmosphere guided many of the recent efforts in solving the initialization problem by "updating" techniques. The discussion above was an attempt to make this idea rigorous to a certain extent by confining the "updating" to an infinitesimal interval as it were, rather than the finite and rather long one over which it is done in numerical experiments with GCMs. At the present stage then it seems plausible to conclude that this idea is not true in the precise sense we tried to define. Consideration involving the solution of the primitive equations over finite time spans appears to lead to a certain confusion in trying to define the concepts involved, as well as to rather inconclusive numerical results which are difficult to interpret.

In the sense of our compatibility conditions, as relations between variables not including time derivatives of the wind field, there would seem to be no balanced state of the atmosphere consistent with the primitive equations. However, if it became practically feasible to measure at least one of the velocity components, or a scalar univalently connected with the wind field (such as the speed or the horizontal wind direction) over a sufficiently fine uniform grid, our compatibility conditions would become sufficient, together with appropriate boundary conditions, in order to determine the initial velocity field uniquely. This would then solve the completeness problem for the initial data of the primitive equation, still leaving open the accuracy problem, which would have to be solved by an estimate similar to (1) of Sec. 2. This estimate would then depend on the (nonlinear) functional of the velocity field actually measured. It is hoped that the present approach will prove fruitful in solving the initialization problem along the lines discussed in this paragraph.

REFERENCES

1. J.G. Charney, The use of the primitive equations of motion in numerical prediction, Tellus, VII (1955), 22-26.

2. J. Charney, M. Halem and R. Jastrow, Use of incomplete historical data to infer the present state of the atmosphere, The Journal of the Atmospheric Sciences, 26 (1969) 1160-1163.

3. M. Ghil. On balance and initialization, Courant Inst. Report IMM-400, 1973; also submitted to the Journal of the Atmospheric Sciences.

4. G.J. Haltiner, Numerical Weather Prediction, John Wiley and Sons, New York, 1971.

5. P. Morel, G. Lefevre and G. Rabreau, On initialization and nonsynoptic data assimilation, Tellus, XXIII (1971), 197-206.

6. T. Nitta, and J.G. Hovermale, A technique of objective analysis and initialization for the primitive forecast equations, Monthly Weather Review, 97, (1969), 652-658.

7. D.L. Williamson and R.E. Dickinson, Periodic updating of meteorological variables, J. Atmos. Sci., 29 (1972), 190-193.

8. D. Williamson and A. Kasahara, Adaptation of meteorological variables forced by updating, J. Atmos, Sci. 28 (1971), 1313-1324.

This research was supported by the Advanced Research Projects Agency of the Department of Defense and was monitored by U.S. Army Research Office --Durham under Grant No. DA-ARO-D-31-124-72-G113.

Courant Institute of Mathematical Sciences,
New York University.

A CARASSO

The Backward Beam Equation and the Numerical Computation of Dissipative Equations Backwards in Time

ABSTRACT

We present an expository survey of the backward beam equation approach in the numerical computation of parabolic equations backwards in time. We discuss linear and nonlinear problems, and we present the details of several numerical experiments on problems where exact solutions are known. Our discussion includes problems with variable coefficients depending on time, as well as recently obtained results on the computation of the final value problem for Burgers' equation.

1. INTRODUCTION

The purpose of this paper is to survey the main results of [2-6], dealing with the development of a new algorithm for the approximate solution of backwards parabolic equations. The exposition is mostly without proofs, and the reader is referred to the original papers for a more detailed discussion. Since this work was begun, a substantial amount of computational experience has been accumulated. The method appears to be a powerful tool. Some computational experiments, undertaken by B.L. Buzbee at the Los Alamos Scientific Laboratories, have not been published. They are mentioned in Section 3. More recently, [6], the method was extended to an interesting example of a nonlinear equation, Burgers' equation, and extensive numerical experiments were carried out on problems for which exact solutions are known. In Section 5, we describe some of the results obtained in [6]. Further applications to nonlinear problems are currently under way.

2. SOME GENERAL REMARKS ON BACKWARDS DISSIPATIVE EQUATIONS

We shall focus attention on a special but quite interesting class of ill-posed problems, namely the problems which arise when the time direction is reversed in a dissipative evolution equation. Such equations distinguish a time direction as they describe irreversible phenomena. While the problem of determining the future from the present is well understood analytically, and a considerable literature exists which deals with the effective numerical computation of such forward problems, the same is not true of the backwards problem. Attempting to reconstruct the past from the present leads to tremendous difficulties, as the solutions depend discontinuously on the data, and this in an essential way. That is, continuous dependence cannot usually be restored by reconsidering the question in some other metrizable linear space. What seems necessary to restore continuity is the imposition of additional constraints on the class of admissible solutions, such as requiring the solutions to be positive or to satisfy an a priori bound. On the other hand, while the constrained problem is well posed in the analytical sense, there remains the question of devising effective numerical methods in which the constraints can be incorporated, if one wishes to approximate the solutions. Thus, the standard stable marching schemes (such as the Crank-Nicolson or other Padé approximants) which have been widely successful in the numerical computation of forward dissipative problems, are necessarily unstable when the time direction is reversed. This is a general theorem to be found in [17, p.59]. What is missing in these classical schemes is a means of incorporating the constraints. In recent years, it has become apparent that the essential difficulties in the above class of ill-posed problems are algorithmic in character. By and large, the numerical analysis of these problems is not as well developed as are the analytical questions such as backwards uniqueness and stability under a prescribed bound. Nevertheless a great deal of effort and ingenuity has been applied towards the effective numerical computation of such problems; we refer to [16] for an extensive bibliography, and to the paper by Miller in the present volume for a viewpoint somewhat different from ours.

It should be pointed out that not all methods which have been proposed are equally effective or applicable. Thus, methods which require _exact_ knowledge of the data are of limited usefulness in applications. Even when an analytical expression is available for the data, round-off error in digitizing that data plays a considerable role and its effect cannot be ignored. The same is true for schemes which require the data and corresponding solution to have a Fourier transform with compact support. As the spatial mesh is refined, extraneous high frequencies are injected into the solution by the rounding process. These high frequencies may be amplified without bound, in arbitrarily small time intervals, as time evolves backwards. Furthermore, if a smoothing process is used at each time step, it may not be possible to decide which parts of the spectrum should be filtered out. Thus, in nonlinear problems, or even in linear problems with variable coefficients, there may be genuine interactions between different frequency bands.
Several methods which have been proposed (for linear problems) begin by recasting the problem in the form of an integral equation, i.e.

$$S(T)u = v \tag{2.1}$$

where v is the given terminal data at time T, u is the desired initial data, and S(T) is the solution operator at time T in the forward analytic problem. Since $[S(T)]^{-1}$ is unbounded, various regularization techniques are then employed to approximate u. The difficulty here is that except in very simple problems, one generally does not know S(T) explicitly. Thus, if the equation has variable coefficients depending on time, one rarely has formulae for the fundamental solution.
It is a happy fact that many stabilized backwards dissipative problems enjoy the property of _Hölder-continuity_ with respect to the data on compact subintervals of (0,T]. Usually the exponent $\mu(t)$ is a function of t which tends to zero as $t \downarrow 0$. As a simple example, in linear problems with a self adjoint operator, one can give a _sharp_ estimate for the error at time t,

corresponding to an error δ at time T, given an a priori bound M, for the initial data. One has in the L^2 norm,

$$\| \epsilon(t) \| \leq 2 M^{\frac{T-t}{T}} \delta^{\frac{t}{T}} , \quad 0 \leq t \leq T. \tag{2.2}$$

Here $\mu(t) = \frac{t}{T}$, and there is "destruction of information" as $t \downarrow 0$. This loss of information is more severe than that which takes place in the usual well-posed linear problems of mathematical physics. On the other hand, the situation is far worse in other classes of stabilized ill-posed problems. Thus, in [9], F. John gives an example where a much weaker type of continuity, "logarithmic continuity", can be shown to actually hold. In that example, the data must be known to an accuracy of $10^{-400,000,000}$ in order to produce an accuracy of 10^{-3} in the corresponding solutions! An important requirement for a numerical method is that it should preserve the Holder dependence inherent in the analytic problem. This requirement is <u>stronger</u> than stability. The latter simply requires that round-off or other errors in the data not be amplified without bound in finite time; however, stability alone may not prevent the scheme from amplifying errors in the data <u>far beyond</u> the theoretical limit set by the estimate (2.2), resulting in an unwarranted loss of precious information. Thus in [8], an interesting example is given of a stable marching scheme for the backwards heat equation, in which logarithmic continuity with respect to the data actually holds. Similarly, [18], Tichonov's method for the backward heat equation is only logarithmically continuous with respect to the data.

Even at the analytical level, there remain genuine difficulties in obtaining a priori stability estimates in backwards dissipative problems. Such estimates are of fundamental importance as they measure the rate at which information is destroyed as time evolves backwards. To illustrate this point, consider the situation for the Navier-Stokes equations backwards in time. In [12] the authors consider the class of smooth solutions of the Navier-Stokes equations, in a space time domain $\Omega \times [0,T]$ satisfying

$$\sup_{(x,t)\in\Omega\times[0,T]} \{|u|^2 + |\text{curl } u|^2 + |u_t|^2\} \le N^2 , \tag{2.3}$$

where N is a given a priori bound. Let ν be the kinematic viscosity, V the volume of Ω, and let $u_1(x,t)$, $u_2(x,t)$ be two smooth solutions of the Navier-Stokes equations, satisfying (2.3) in $\Omega\times[0,T]$, and such that,

$$\|u_1(\cdot, T) - u_2(\cdot, T)\|^2_{L^2(\Omega)} \le \delta. \tag{2.4}$$

It is shown in [12] that then, for $0 \le t \le T$,

$$\|u_1(\cdot,t) - u_2(\cdot,t)\|^2_{L^2(\Omega)} \le (4N^2)^{1-\mu(t)}\delta^{\mu(t)} \exp\left[\frac{N^4(t-\mu(t)T)}{\nu^2}\right] \tag{2.5}$$

where $\mu(t)$ is given by

$$\mu(t) = \frac{1-\exp\left[\frac{2(N^2+1)t}{\nu}\right]}{1-\exp\left[\frac{2(N^2+1)T}{\nu}\right]} . \tag{2.6}$$

The estimate (2.5) establishes <u>Hölder-continuous</u> dependence on the data and implies backwards uniqueness of smooth solutions. On the other hand, as far as computing the solutions are concerned, (2.5) is rather disconcerting. For example, if $V=T=N=1$, $\nu=10^{-1}$, and $\delta=10^{-50}$, we have from (2.5) at $t=T/2=1/2$,

$$\|u_1(\cdot,1/2)-u_2(\cdot,1/2\|^2_{L^2(\Omega)} \le 4e^{50}(10^{-50})^{e^{-20}} \approx 10^{22}. \tag{2.7}$$

Moreover, (2.7) is little changed by choosing $\delta=10^{-500}$, since the rate at which $\mu(t)\downarrow 0$ is so large, even at such small Reynolds numbers. It is not known whether the exponent $\mu(t)$ is sharp, or whether constraints different from (2.3), possibly involving other combinations or derivatives, would result in a more encouraging estimate. In non-linear problems, a major task appears to be that of isolating classes of equations for which one can obtain fairly reasonable Hölder estimates. As a very small beginning, the one dimensional Burgers' equation,

$$u_t = \nu u_{xx} - uu_x + f(x,t), \quad 0 \le x \le L, \quad 0 \le t \le T, \tag{2.8}$$

is considered in [6]. It is straightforward to show that if $u_1(x,t)$, $u_2(x,t)$, are two solutions satisfying

$$\text{Max}\ \{|u_i|,\ |u_{i,t}|,\ |u_{i,tt}|,\ |u_{i,xt}|\} \le N,\ i=1,2, \tag{2.9}$$

for $(x,t) \in [0,L]\times[0,T]$, and if

$$\|u_1(\cdot,T) - u_2(\cdot,T)\|_{L^2} \le \delta, \tag{2.10}$$

then for $0 \le t \le T$,

$$\|u_1(\cdot,t) - u_2(\cdot,t)\|_{L^2} \le 2\ K(t)\ N^{\frac{T-t}{T}}\ \delta^{\frac{t}{T}}, \tag{2.11}$$

where

$$K(t) = \exp\left[\frac{4NL+t(T-t)\{(NL)^2+(1+3\nu)NL\}}{4\nu}\right]. \tag{2.12}$$

In [6], extensive numerical experiments are presented for the final value problem for Burgers' equation. Some of these results will be described in the present paper later on. (See Section 5). One of the points suggested by these experiments is that the factor $K(t)$, which involves the Reynolds number, appears to play a significant role only when the solutions to Burgers' equation develop steep gradients approaching a "shock". For more reasonable solutions, one finds that, even in single precision, i.e. with a unit round-off error of about 10^{-8}, one can often attain significant accuracy at 90% of the way back from T=1. Thus, the difficulty of reconstructing past steep gradients from future smoothed data appears to be explained by the factor $K(t)$ in (2.11). Conceivably, in the case of the Navier-Stokes equations, the similar exponential factor in (2.5) may alone suffice to account for the difficulty of reconstructing steep gradients, and in some suitable norm, there might well exist a Hölder estimate in which $\mu(t)$ is independent of the Reynolds number, and decays linearly with t.

3. THE BACKWARD BEAM EQUATION APPROACH IN SELF-ADJOINT PROBLEMS WITH TIME INDEPENDENT COEFFICIENTS

We shall now describe a new method, which was recently developed in [2], for computing linear self-adjoint parabolic equations backwards in time. In this section we consider the case where the spatial operator, A, is independent of t. The case where A depends on t is more subtle and is discussed in Section 4.

Let f(x) be a given function in $L^2(\Omega)$ where Ω is a bounded domain in R^N, in R^N with a smooth boundary $\partial\Omega$. Let A be a non-negative self-adjoint operator in $L^2(\Omega)$; in the concrete cases, A is the unbounded operator corresponding to a self-adjoint elliptic boundary value problem in Ω, with, say, zero Dirichlet data on $\partial\Omega$. Given the positive constants δ, M, T, we consider the following problem.

Find all solutions of

$$u_t = -Au, \quad 0 < t \leq T, \tag{3.1}$$

such that

$$\|u(\cdot,T) - f\| \leq \delta , \tag{3.2}$$

and

$$\|u(\cdot,0)\| \leq M. \tag{3.3}$$

To solve this stabilized backwards problem, consider the following device.

Set

$$k = \frac{1}{T} \log \left(\frac{M}{\delta}\right) \tag{3.4}$$

and then put $v = e^{kt}u$ in (3.1). This leads to

$$v_t = -(A-k)v, \quad 0 < t \leq T, \tag{3.5}$$

$$\|v(\cdot,T) - e^{kT}f\| \leq e^{kT}\delta, \qquad \|v(0)\| \leq M. \tag{3.6}$$

Differentiating (3.5) with respect to t, we obtain the "backward beam equation" associated with (3.1), namely,

$$v_{tt} = Bv, \quad 0 < t \leq T, \tag{3.7}$$

where $B = (A-k)^2$. Thus, B is a non-negative self-adjoint operator in $L^2(\Omega)$; in particular, B is "m-accretive". For such equations, it is easy to show that solutions are norm-convex. We have,

$$\frac{d^2}{dt^2}\|v(t)\|^2 = 2\|v'(t)\|^2 + 2\,\mathrm{Re}(Bv, v) \geq 0. \tag{3.8}$$

In particular, if v(t) is a solution of (3.7),

$$\|v(t)\| \leq \frac{T-t}{T}\|v(0)\| + \frac{t}{T}\|v(T)\| . \tag{3.9}$$

The last inequality suggests that the "initial-terminal value" or "two-point" problem is well-posed for (3.7), i.e. data should be prescribed at t=0 and at t=T. Using Hadamard's classical example of the Cauchy problem for Laplace's equation, it is easily shown that the initial value problem is in general ill-posed for (3.7). It also follows from the spectral representation of B, that solutions to the two point problem exist, for <u>arbitrary</u> data v(0), v(T) in $L^2(\Omega)$; moreover, as shown in [5], (3.7) has the "smoothing" property. That is, if A has sufficiently smooth coefficients, arbitrarily high Sobolev norms of the solution at time t, $0 < t < T$, can be estimated in terms of the L^2 norms of the data v(0), v(T). In a very real sense, (3.7) is an "elliptic" equation in Hilbert space.

Let $w(\cdot,t)$ be the unique solution of

$$w_{tt} = Bw, \quad 0 < t < T, \tag{3.10}$$

$$w(0) = 0 \quad w(T) = e^{kT}f. \tag{3.11}$$

We then have

Theorem 3.1

Let $u(\cdot,t)$ be any solution of the stabilized backwards problem (3.1)-(3.3). Let k be as in (3.4), and let $w(\cdot,t)$ be the unique solution of (3.10)-(3.11). Then,

$$\|e^{-kt}w(\cdot,t) - u(\cdot,t)\| \le M^{\frac{T-t}{T}} \delta^{\frac{t}{T}} \tag{3.12}$$

Moreover, if A has smooth coefficients, N is the dimension of Ω, q is a positive integer, and $2\sigma > \frac{N}{2} + q$, there is a constant K such that for $0 < t < T$,

$$\max_{|\beta|\le q} \|D^\beta u(\cdot,t) - e^{-kt}D^\beta w(\cdot,t)\|_\infty \le \tag{3.13}$$

$$K\left\{(t)^{-\sigma} + (T-t)^{-\sigma} + \left(\frac{\log(\frac{M}{\delta})}{T}\right)^{\sigma}\right\} M^{\frac{T-t}{T}} \delta^{\frac{t}{T}}.$$

Proof.

If u is any solution of (3.1)-(3.3), $v = e^{kt}u$ satisfies (3.7) and the inequalities (3.6). Let $z(\cdot,t) = v - w$. Then, from (3.9) and (3.6), we have,

$$\|e^{-kt}z(\cdot,t)\| \le \frac{T-t}{T} Me^{-kt} + \frac{t}{T} e^{k(T-t)}\delta = M^{\frac{T-t}{T}} \delta^{\frac{t}{T}}. \tag{3.14}$$

This proves (3.12). The proof of (3.13) is more complicated and we refer the reader to [5].

Remark 1. Since the estimate (2.2) is sharp, for the difference of any two solutions of the backwards problem, it follows that $e^{-kt}w$ above, is a "best-possible" L^2 approximate to any solution of the backwards problem. Moreover, even though the data $f(x)$ is an approximation to $u(\cdot,T)$ in the L^2-norm, $e^{-kt}w$

approximates the solutions of the backwards problem, together with their derivatives, in the L^∞ norm, on $0 < t < T$.

Remark 2. By using (3.12) and the triangle inequality, we obtain an independent proof of the convexity estimate, (2.2), for the difference of any two solutions of the backwards problem. Similarly, (3.13) and the triangle inequality lead to a maximum norm stability estimate, for the derivatives of any two solutions, in terms of the L^2 norm of the data.

In actual numerical computation of the solution of (3.10)-(3.11), the time variable is discretized using a centered time discretization. With $T = (N+1)\Delta t$, we have,

$$\frac{w^{n+1}-2w^{n}+w^{n-1}}{\Delta t^{2}} = Bw^{n}, \quad n=1,2,\ldots,N, \tag{3.15}$$

$$w^{0} = 0, \qquad w^{N+1} = e^{kT}f. \tag{3.16}$$

This system of linear equations can be written in tridiagonal matrix form, with an unbounded operator along the main diagonal. See [3], [4], [2]. Using the tridiagonal algorithm, one obtains the existence and uniqueness of solutions to (3.15), (3.16), together with the basic normconvexity property (3.9), for the solution of this finite difference analog; the proof uses only the "m-accretiveness" of B. Moreover, this method of proof provides one of several possible algorithms for the solution of this system of linear equations. To discretize the spatial operator B, one uses finite difference analogs of the elliptic operator, or Galerkin methods using trial functions satisfying the boundary conditions. Either method preserves the "m-accretiveness" of B. Consequently, the fully-discrete scheme is unconditionally stable and has the property (3.9). Finally, estimates such as (3.12), (3.13), supplemented by the truncation error at time t in the fully-discrete scheme, remain valid for the solution of the fully -discrete problem. We refer the reader to [2] for a more detailed discussion of these matters. By and large, the numerical analysis of (3.10), (3.11), is very similar to that for elliptic boundary value problems, for which an abundant literature exists. Thus, direct or iterative methods may be used to solve the system of linear equations. In [2], a computational example is discussed in detail for a one dimensional problem. Further examples will be given later in the present paper, when we discuss the backwards problem for Burgers' equation (See Section 5). Finally, we mention some successful computations, on <u>two-dimensional</u> problems in rectangular regions, carried out by B.L. Buzbee at the Los Alamos Scientific Laboratories.

4. SELF-ADJOINT PROBLEMS WITH TIME DEPENDENT COEFFICIENTS

An important feature of the backward beam equation approach is that the method is applicable to self-adjoint parabolic equations with smooth time dependent coefficients. The assumption of smooth dependence on time is important, as backwards uniqueness may fail for non-smooth coefficients. See the example of Miller in [14].

For each $t \geq 0$, let $a(t;u,v)$ be the symmetric bilinear form on $H_0^m(\Omega)$ given by

$$a(t;u,v) = \sum_{|p|,|q| \leq m} \int_\Omega a_{pq}(x,t) D^q u \overline{D^p v} \, dx \tag{4.1}$$

where the a_{pq} depend smoothly on x and t, and

$$a_{pq} = \bar{a}_{qp}. \tag{4.2}$$

We assume $a(t;u,v)$ to be strongly coercive on $H_0^m(\Omega)$ i.e. there exists a positive constant ω, independent of t, such that

$$a(t;v,v) \geq \omega \| v \|_m^2, \qquad v \in H_0^m(\Omega), \tag{4.3}$$

where $\| \ \|_m$ denotes the Sobolev norm. Let $a(t;u,v)$ be the bilinear form obtained from $a(t;u,v)$ by replacing a_{pq} by $\dot{a}_{pq} = \frac{\partial}{\partial t} a_{pq}$. The form $a(t;u,v)$ will play an important role in the subsequent discussion. Let $P(t)$ be the unbounded self-adjoint operator in $L^2(\Omega)$ defined by $a(t;u,v)$ i.e,

$$\langle P(t)v,v\rangle = a(t;v,v), \forall v \in H_0^m(\Omega) \tag{4.4}$$

where, $\langle \ , \ \rangle$ is the scalar product in $L^2(\Omega)$. An integration by parts shows that $P(t)$ corresponds to the self-adjoint elliptic partial differential operator,

$$P_0(t) = \sum_{|p|,|q|\leq m} (-1)^{|p|} D^p(a_{pq}(x,t)D^q u), \; x \in \Omega, \; t > 0, \tag{4.5}$$

together with the Dirichlet boundary conditions

$$D^p u=0 \quad \text{on } \partial\Omega \;, \quad |p| \leq m - 1, \; t \geq 0. \tag{4.6}$$

We shall consider the parabolic problem

$$u_t = -P(t)u, \; t > 0. \tag{4.7}$$

and we note that from (4.6), the domain of $P(t)$ is <u>fixed</u> as t varies. We now introduce the following definitions.

<u>Definition</u>

The parabolic problem (4.7) is <u>minimal-smoothing</u> on $[0,T]$ if

$$\dot{a}(t;v,v) \leq 0 \quad \forall \quad v \in V, \quad 0 \leq t \leq T. \tag{4.8}$$

It is <u>strongly-smoothing</u> if

$$\dot{a}(t;v,v) \leq 2\gamma \,\|v\|^2 \;, \quad \gamma > 0. \tag{4.9}$$

where $\| \ \|$ denotes the L^2 norm. We say that (4.7) is <u>maximal-smoothing</u> if

$$\dot{a}(t;v,v) \leq \alpha \,\|v\|_m^2, \quad \alpha > 0. \tag{4.10}$$

The above definition distinguishes three broad classes of problems. Further refinements are clearly possible. In the interest of simplicity of exposition, these refinements are not considered here. It will be seen later that the maximal-smoothing case is the hardest to compute backwards in time. This difficulty is not a defect of our method. Rather, it is associated with the type of Hölder estimate which obtains in that case. Each of the maximal-smoothing and strongly-smoothing cases can be reduced to the minimal-smoothing case by means of a preliminary transformation.

In the maximal-smoothing case this reduction is accomplished by stretching the time variable. With ω and α the constants in (4.3) and (4.10), define the function

$$\psi(s) = \left(\frac{\omega}{\alpha}\right)\log(1+\alpha s/\omega), \quad s \geq 0. \tag{4.11}$$

Then, $\psi'(s) > 0$, $\psi''(s) < 0$ and

$$\omega\psi'' + \alpha(\psi')^2 = 0. \tag{4.12}$$

From the bilinear form $a(t;u,v)$ in (4.1), we construct the form $b(s;u,v)$ where

$$b(s;u,v) = a(\psi(s);u,v)\psi'(s), \quad s \geq 0, \tag{4.13}$$

$$= \sum_{|p|,|q|\leq m} \int_\Omega b_{pq}(x,s)D^q u\, \overline{D^p v}\, dx,$$

with

$$b_{pq}(x,s) = a_{pq}(x,\psi(s))\psi'(s), \quad s \geq 0. \tag{4.14}$$

Let $b'(s;u,v)$ be the symmetric bilinear form on $H_0^m(\Omega)$, obtained from (4.13) by replacing $b_{pq}(x,s)$ by $\frac{\partial b_{pq}}{\partial s}$. We then have,

$$b'(s;u,v) = \psi''\, a(\psi(s);u,v) + (\psi')^2\dot{a}(\psi(s);u,v). \tag{4.15}$$

Hence, using (4.3), (4.10), (4.12) and the fact that $\psi'' < 0$, we obtain from (4.15),

$$b'(s;v,v) \leq [\alpha(\psi')^2 + \omega\psi'']\,\|v\|_m^2 = 0, \quad \forall\, v \in H_0^m(\Omega). \tag{4.16}$$

We now put $t = \psi(s)$ in the parabolic problem (4.7). Let

$$\xi(x,s) = u(x,\psi(s)), \quad x \in \Omega, \quad s \geq 0; \tag{4.17}$$

then, ξ satisfies the parabolic problem,

$$\xi_s = -G_0(s)\xi, \quad x \in \Omega, \ s > 0, \tag{4.18}$$

$$D^p\xi = 0, \ x \in \partial\Omega, \ s \geq 0, \quad |p| \leq m-1, \tag{4.19}$$

where $G_0(s) = \psi'(s)\, P_0(\psi(s))$. This transformed parabolic problem is the one generated by the symmetric bilinear form $b(s;u,v)$. Since $b'(s;v,v) \leq 0$, we have the minimal-smoothing case for the transformed problem.

In the <u>strongly smoothing</u> case, we put

$$b(t;u,v) = a(t;u,v) - 2\gamma t \int_\Omega u\,\overline{v}\,dx. \tag{4.20}$$

Then, from (4.9), $b(t;v,v) \leq 0$. Next, put

$$\xi(x,t) = e^{\gamma t^2} u(x,t), \ x \in \Omega, \ t > 0 \tag{4.21}$$

in the parabolic problem (4.7). Then ξ satisfies,

$$\xi_t = -G_0(t)\xi, \ x \in \Omega, \ t > 0, \tag{4.22}$$

$$D^p\xi = 0, \ x \in \partial\Omega, \quad t \geq 0, \quad |p| \leq m - 1, \tag{4.23}$$

where $G_0(t) = P_0(t) - 2\gamma t$. This is the problem generated by the bilinear form $b(t;u,v)$ in (4.20). Thus (4.21), transforms the strongly smoothing case into the minimally smoothing case.

For the purposes of the following discussion , it may now be assumed without loss of generality that the parabolic problem,

$$u_t = -P(t)u, \quad t > 0. \tag{4.24}$$

is minimally smoothing on the interval [0,T]. Given $f(x)$ in $L^2(\Omega)$, and the positive constants δ, M,T, consider the following problem. Find all solutions of (4.24) such that

$$\|u(\cdot,T) - f\| \leq \tilde{\delta} \tag{4.25}$$

$$\|u(\cdot,0)\| \leq M . \tag{4.26}$$

As in Section 3, we put $\tilde{k} = \frac{1}{T} \log \left(\frac{M}{\tilde{\delta}}\right)$ and $v = e^{\tilde{k}t} u$ in (4.24), to obtain

$$v_t = -(P(t) - \tilde{k})v, \quad 0 < t < T. \tag{4.27}$$

Differentiating with respect to t, we obtain

$$v_{tt} = A(t)v, \quad 0 < t < T, \tag{4.28}$$

where $A(t)$ is an unbounded self-adjoint operator in $L^2(\Omega)$. See [4, Section 3]. In fact, for each fixed $t > 0$, $A(t)$ is the unbounded operator corresponding to the following elliptic boundary value problem of order 4m in Ω:

$$[(P_0(t) - \tilde{k})^2 - \dot{P}_0(t)]u = 0, \ x \in \Omega, \tag{4.29}$$

$$D^p u = D^p[P_0(t)u] = 0, \quad x \in \partial\Omega, \quad |p| \leq m - 1; \tag{4.30}$$

where $\dot{P}_0(t)$ is the differential operator obtained from $P_0(t)$ in (4.5) by differentiating the coefficients with respect to t. Since the problem (4.24) is minimally-smoothing by hypothesis, we have,

$$\langle \dot{P}_0(t)v, v\rangle \leq 0, \ \forall \ v \in H_0^m(\Omega). \tag{4.31}$$

Consequently, $A(t)$ in (4.28) is a <u>non-negative</u> self-adjoint operator for each t. Note, however, that even though (4.24) is a <u>fixed</u> domain parabolic problem, the backward beam equation (4.28) now involves a <u>variable</u> domain operator, $A(t)$, in general. See [4, Section 3]. Unlike the problem in Section 3, it is no longer possible to prove existence theorems for (4.28) by using the spectral representation of $A(t)$ at each t. Remarkably enough, one can obtain strong results, (i.e. existence of solutions lying in the domain of $A(t)$ for each t), even in this variable domain case, for the two-point problem associated with the finite difference analog of (4.28). No assumptions need be made about the manner in which $D_A(t)$ varies with t, and the proof uses only the "m-accretiveness" of A(t) for each t. Let $T = (N+1)\Delta t$, and consider the system of difference equations,

$$\frac{w^{n+1} - 2w^n + w^{n-1}}{\Delta t^2} - A^n w^n = 0, \quad n=1,2\ldots,N, \tag{4.32}$$

$$w^0 = a, \qquad w^{N+1} = b, \tag{4.33}$$

where $a,b \in L^2(\Omega)$, and $A^n \equiv A(n\Delta t)$. In [3], the following theorem is proved.

Theorem 4.1

There exists a unique solution, w(t), in (4.32), (4.33), with $w(t) \in D_A(t)$, for each $t = n\Delta t$, $n = 1,2,\ldots,N$, and this for arbitrary $a,b \in L^2(\Omega)$. Moreover,

$$\|w(t)\| \leq \frac{T-t}{T}\|a\| + \frac{t}{T}\|b\| . \tag{4.34}$$

The proof of Theorem 4.1 given in [3] is constructive, and is based on the tridiagonal algorithm. If finite element methods are used to discretize A(t) for each t, such methods preserve the accretiveness of A(t). Consequently, the resulting fully-discrete scheme also has the norm convexity property (4.34). In practice, iterative methods, such as block relaxation techniques, may be used to solve the block tridiagonal system of linear equations. See [4].

We now turn to the question of approximating the solutions to the parabolic problem (4.7), backwards in time, given an a priori bound, M, for the initial data, and given a function $f(x) \in L^2(\Omega)$, such that $\|u(\cdot,T) - f\| \leq \delta$. If (4.7) is minimally-smoothing, we solve the system (4.32) with the two-point conditions,

$$w^0 = 0, \quad w^{N+1} = e^{kT}f, \tag{4.35}$$

and

$$k = \frac{1}{T} \log \frac{M}{\delta} . \tag{4.36}$$

We then define,

$$u_{app}(t) = e^{-kt} w(t) \tag{4.37}$$

as our approximation to the solutions of the backwards problem. Thus, this case is treated in exactly the same way as the problem in Section 3.

If (4.7) is strongly-smoothing, we first transform to the minimal case by means of (4.21), and solve the resulting system (4.32) for the transformed problem, with the two-point conditions,

$$w^0 = 0, \quad w^{N+1} = e^{\gamma T^2} e^{kT} f, \tag{4.38}$$

where

$$k = \frac{1}{T} \{\log M - \log(e^{\gamma T^2} \delta)\}. \tag{4.39}$$

We then define,

$$u_{app}(t) = e^{-\gamma t^2} e^{-kt} w(t) \tag{4.40}$$

as our approximation to the solutions of the original problem. Finally, in the maximal-smoothing case, we transform to the stretched variable, s, as in (4.17). Let $\Delta s = S/N+1$, where

$$S = \left(\frac{\omega}{\alpha}\right) [e^{\alpha T/\omega} - 1]. \tag{4.41}$$

We solve the system (4.32) in the s-variable with the two-point conditions,

$$w^0 = 0, \quad w^{N+1} = e^{kS} f \tag{4.42}$$

and $k = \frac{1}{S} \log \left(\frac{M}{\delta}\right)$.

Let

$$\psi^{-1}(t) = \frac{\omega}{\alpha} [e^{\alpha t/\omega} - 1]. \tag{4.43}$$

As our approximation to the solutions of the original problem, we define,

$$u_{app}(t) = \exp[-k\psi^{-1}(t)]w(\psi^{-1})t)), \tag{4.44}$$

where w(s) is the solution of (4.32), (4.42). We then have the following result.

<u>Theorem 4.2</u>

Let u(t) be any solution of the stabilized backwards problem for (4.7), and consider $u_{app}(t)$. In the <u>minimally-smoothing</u> case, we have,

$$||u(t) - u_{app}(t)|| \leq M^{\frac{T-t}{T}} \delta^{\frac{t}{T}} + O(\Delta t^2).$$

In the <u>strongly-smoothing</u> case,

$$||u(t) - u_{app}(t)|| \leq \exp[\gamma t(T-t)]M^{\frac{T-t}{T}} \delta^{\frac{t}{T}} + O(\Delta t^2). \tag{4.46}$$

Finally, in the <u>maximal-smoothing</u> case,

$$||u(t) - u_{app}(t)|| \leq M^{1-\mu(t)}\delta^{\mu(t)} + O(\Delta t^2), \tag{4.47}$$

where

$$\mu(t) = \frac{e^{\alpha t/\omega} - 1}{e^{\alpha T/\omega} - 1} . \tag{4.48}$$

<u>Proof</u>

The proof of each of the three inequalities follows from the norm-convex property of the solution of the backward beam equation associated with each of the transformed problems, as in the proof of (3.12) in Theorem 3.1, together with a subsequent inverse transformation in the case if the last two inequalities. See [2]. The extra term $O(\Delta t^2)$ represents the combined spatial and time discretization errors, as we are now considering the fully discrete scheme as opposed to the evolution equation (4.28). It is assumed that the spatial mesh is chosen so that the spatial truncation error is of the same magnitude as that of the time discretization.

Remark By making $\Delta t \to 0$ in the above error estimates, we recover logarithmic convexity estimates originally obtained by Agmon-Nirenberg, [1]. Note the exponential decay to zero as $t \downarrow 0$, of the exponent $\mu(t)$ in (4.47). An example of the maximal-smoothing case is provided by a simple diffusion equation in which the diffusion coefficient grows with time. The strongly smoothing case corresponds to a diffusion equation with a growing zero order term. The minimal case corresponds to a constant or decaying diffusion coefficient. By considering a diffusion coefficient depending only on t, one can show that (4.47) is sharp. In such a problem, considerably more precision in measurement is necessary at time T, in order to attain significant accuracy backwards in time, as compared with the other two cases.

5. COMPUTING SMALL SOLUTIONS OF BURGERS' EQUATION BACKWARDS IN TIME

We shall now describe some recent results dealing with the application of the backward beam method in the computation of the final value problem for Burgers' equation. The reader is referred to [6], for proofs and a more detailed discussion of the main results. We consider the following initial boundary value problem for the one dimensional Burgers' equation,

$$u_t = \nu u_{xx} - uu_x + f(x,t), \quad 0 \leq x \leq L, \quad 0 \leq t \leq T, \tag{5.1}$$

$$u(0,t) = u(L,t) = 0, \quad t \geq 0, \tag{5.2}$$

$$u(x,0) = a(x), \quad 0 \leq x \leq L, \tag{5.3}$$

where $a(x)$ and $f(x,t)$ are sufficiently smooth that the (unique) solution of (5.1)-(5.3) has sufficiently many derivatives on $[0,T]$. Let A be the positive self-adjoint operator corresponding to $-u''$ with zero boundary conditions, and let

$$F(u) = -uu_x. \tag{5.4}$$

We may then write (5.1)-(5.3) in the form of an evolution equation in $L^2[0,L]$, viz,

$$u_t = -\nu A u + F(u) + f(t), \quad 0 < t < T, \tag{5.5}$$

$$u(0) = a. \tag{5.6}$$

One way of solving the <u>forward</u> problem is by means of the following iterative procedure,

$$u_t^0 = -\nu A u^0 + f(t), \quad 0 < t < T, \tag{5.7}$$

$$u^0(0) = a, \tag{5.8}$$

and for each m=1,2,3,...,

$$u_t^m = -\nu A u^m + F(u^{m-1}) + f(t), \quad 0 < t < T, \tag{5.9}$$

$$u^m(0) = a. \tag{5.10}$$

In fact, this procedure was used by Kato and Fujita in [10], [11], as a means of proving existence and uniqueness theorems for the Navier-Stokes equations, which they viewed as an initial value problem in Hilbert space. An important feature of the above iteration is that one proceeds through a sequence of inhomogeneous linear parabolic problems with <u>constant</u> coefficients. Concerning the convergence of this iterative process, we have the following theorem. See [6].

<u>Theorem 5.1</u>

Let $a(x)$ belong to $D(A^{\frac{1}{2}})$, and let $f(t) \in D(A^{\frac{1}{2}})$ with $\|A^{\frac{1}{2}} f(t)\| \in L^1[0,T]$. Let

$$\left(\frac{64\ LT}{\nu}\right)^{\frac{1}{2}} \left[\|A^{\frac{1}{2}} a\| + \int_0^T \|A^{\frac{1}{2}} f(s)\| ds \right] < 1 \; ; \tag{5.11}$$

then $A^{\frac{1}{2}} u^m$ exists on $[0,T]$ for every m. Let $\theta = 1 - [1 - (\frac{64LT}{\nu})^{\frac{1}{2}} ||| A^{\frac{1}{2}} u^0 |||\,]^{\frac{1}{2}}$. Then $0 < \theta < 1$, and

$$||| A^{\frac{1}{2}} u^m ||| \leq \theta \left(\frac{\nu}{16\ LT}\right)^{\frac{1}{2}}. \tag{5.12}$$

Moreover, if u(t) is the unique solution of (5.5), (5.6),

$$||| A^{\frac{1}{2}}(u^m-u) ||| \leq \theta^{m+2}\left(\frac{\nu}{64\ LT}\right)^{\frac{1}{2}} . \qquad (5.13)$$

In the above statement of the theorem, $||| v |||$ denotes

$\sup_{0\leq t\leq T} \| v(t) \|$, and $\| A^{\frac{1}{2}}v(t) \|^2 = \int_0^L |v_x(x,t)|^2 dx$.

The condition (5.11) is sufficient for convergence in the norm $|||\ \ |||$. In practice, convergence may occur even if (5.11) is severely violated, by as much as a factor of 1000 in some cases. See [6] for several numerical examples. On the other hand, (5.11) cannot be too severely violated for convergence to occur on [0,T]. In general, the convergence of (5.7)-(5.10) is only local in time, even if a unique smooth solution exists for all t > 0. An example of divergence, except for t sufficiently small, is given in [6]. In that example, the exact solution is known and is found to develop steep gradients, i.e. a smooth approximation to a "shock" evolves from the initial data. Convergence occurs before the gradients become too steep. See also Example 2 and Figure 1 below.

The essential idea behind the algorithm for the backwards problem, is to solve each linear parabolic problem in the Kato-Fujita sequence of iterates, backwards in time, using the value of the solution to (5.5), (5.6) at time T. Each such linear backwards problem is solved via the backward beam equation discussed in Section 3. Since each linear problem has constant coefficients, the "Fourier Method" of Kreiss-Oliger can be used to great advantage. That is, one calculates the spatial derivatives at the mesh points, by differentiating the trigonometric polynomial which interpolates the function values at equally spaced grid points, using at least 2 points per significant wave length. If the data, inhomogeneous term, and solution have smooth periodic extensions, this technique for differentiation is highly accurate as shown in [13]. It is also extremely attractive in that one can make use of Fast Fourier Transform algorithms. (Such a method has been employed by

Orszag in [15], for the Navier-Stokes equations.) Finally, with this particular technique for discretizing the space variable, the algebraic problem of inverting the "block tridiagonal" matrix in (3.15) becomes trivial. One has a scalar positive definite tridiagonal matrix to invert for each Fourier component in turn, and such matrices can be efficiently inverted by a standard algorithm. It should be remarked that the high accuracy in spatial discretization is quite important in the present equation, where there may be high frequency components in the solution at positive times, even if such components are absent in the initial data. This is a basic property of the homogeneous equation. See the classic paper by Julian Cole in [7].

To motivate our main result, consider the following "thought experiment". Let $\{u^m(t)\}_{m=0}^{\infty}$ be the sequence of iterates in the forward problem, and let $\underset{\sim}{u}^m(T) = b^m$. Imagine that $b_m(x)$ is known approximately. Let $\tilde{b}_m$ be the approximate value of b_m, and suppose

$$\| A^{\frac{1}{2}}(b_m - \tilde{b}_m) \| \leq \delta, \quad m = 0,1,2,\ldots ; \tag{5.14}$$

Suppose further that (5.11) is satisfied for the forward iteration. Then, from (5.12), we have

$$||| A^{\frac{1}{2}}u^m ||| \leq M = \theta\left(\frac{\nu}{64\ LT}\right)^{\frac{1}{2}}. \tag{5.15}$$

We may then pose the following linear backwards parabolic problem for each iterate $u^m(t)$:

Find all solutions of

$$u_t^m = -\nu A u^m + F(u^{m-1}) + f(t), \quad 0 < t \leq T, \tag{5.16}$$

such that

$$\| A^{\frac{1}{2}}(\tilde{b}_m - u^m(T)) \| \leq \delta \tag{5.17}$$

$$\| A^{\frac{1}{2}}u^m(0) \| \leq M. \tag{5.18}$$

Using the backward beam equation with $k = \frac{1}{T} \log (\frac{M}{\delta})$, each $u^m(t)$ may then be approximately solved backwards in time, and used to generate a new approximate inhomogeneous term for the next iteration. Because of the "destruction of information" as $t \downarrow 0$, it is clear that sizable errors will be generated at $t = 0$ and passed on to the next successive iterate. The next theorem assesses the accumulated error after m steps of this process. See [6] for the proof.

<u>Theorem 5.2</u>

Let (5.11) be satisfied. Let $w^m(t)$ be the sequence of successive approximations obtained via the backward beam equation. Let $u(t)$ be the solution of (5.5), (5.6). Then there exists a positive constant C_m, depending only on m, such that

$$||A^{\frac{1}{2}}w^m(t) - A^{\frac{1}{2}}u(t)|| \leq C_m [\log(\frac{M}{\delta})]^{\beta_m} M^{\frac{T-t}{T}} \delta^{\frac{t}{T}} + \theta^{m+2}(\frac{\nu}{64LT})^{\frac{1}{2}} \tag{5.19}$$

where $\beta_m = 2^{m-3} + \frac{1}{2}$.

<u>Remark</u> It follows from (5.19) that given any $\epsilon > 0$, one can make $||A^{\frac{1}{2}}w^m(t) - A^{\frac{1}{2}}u(t)|| < \epsilon$, uniformly on compact subintervals of $(0,T]$, by choosing δ sufficiently small and m sufficiently large. On the other hand, even though (5.11) is satisfied so that the forward iterates, $u^m(t)$, converge on $[0,T]$, the above inequality does not imply convergence, for fixed $\delta > 0$, as $m \to \infty$.

The above theorem is not strictly applicable to the algorithm which is used in practice. In the first place, (5.11) may be severely violated. In the second place, the functions $\tilde{b}_m(x)$ are not available. Rather, one has an approximation $\tilde{b}(x)$ to the terminal value $u(x,T)$, where u is the solution of the non-linear problem (5.1)-(5.3), and,

$$||A^{\frac{1}{2}}u(\cdot,T) - A^{\frac{1}{2}}\tilde{b}|| \leq \delta . \tag{5.20}$$

From physical or other considerations, one may obtain an a priori bound,

$$|||\, A^{\frac{1}{2}}u\, ||| \leq M. \tag{5.21}$$

Using the given data $\tilde{b}(x)$ and setting $k = \frac{1}{T}\log(\frac{M}{\delta})$, each inhomogeneous linear parabolic problem is then solved backwards with the backward beam equation. In an extensive series of numerical experiments, with problems for which exact solutions are known, it is then found that Theorem 5.2 is <u>qualitatively</u> correct as far as the algorithm which is used in practice is concerned. Thus, the solutions are well approximated after a relatively small number of iterations. Further iterations may lead to rapid <u>divergence</u>, even though the forward iteration converges. In many cases, one observes considerable improvement before the onset of divergence. The situation is analogous to that of the divergent power series in the theory of asymptotic expansions, where the first few terms often provide excellent approximations. Furthermore, in several experiments, the distance back into the past where significant accuracy can be attained, is greater than might be expected from the a priori stability estimate for Burgers' equation given in Section 2. We shall now give several numerical examples.

<u>Example 1</u>

Consider

$$u_t = \nu u_{xx} - uu_x, \quad 0 < x < \pi, \quad 0 < t < 1, \tag{5.22}$$

$$u(0,t) = u(\pi,t) = 0, \qquad 0 \leq t \leq 1, \tag{5.23}$$

$$u(x,0) = u_0 \operatorname{Sin} x, \qquad 0 \leq x \leq \pi. \tag{5.24}$$

For any positive ν and any u_0, the exact solution of this problem was obtained by Cole in [7]. It is given by

$$u(x,t) = \frac{4\nu \sum_{n=1}^{\infty} e^{-\nu n^2 t}\, n\, I_n(\frac{u_0}{2\nu}) \operatorname{Sin} n\, x}{I_0(\frac{u_0}{2\nu}) + 2\sum_{n=1}^{\infty} e^{-\nu n^2 t}\, I_n(\frac{u_0}{2\nu}) \operatorname{Cos} n\, x} \tag{5.25}$$

where $I_n(z)$ is the modified Bessel function of the first kind. From (5.25) we observe the effect of non-linearity, in that the initial pure sine wave evolves into a periodic function in which all frequencies are present. It is instructive to associate a Reynolds number with the above problem. Following Cole, [7], we define

$$Re = \frac{u_0 \pi}{\nu} \tag{5.26}$$

In this example we chose $u_0 = 1$ and $\nu = .0025$, so that Re = 126. As far as the sufficient condition for convergence of the forward iteration is concerned, we actually have, in the present case,

$$\left(\frac{64LT}{\nu}\right)^{\frac{1}{2}} \| A^{\frac{1}{2}} a \| \approx 36. \tag{5.27}$$

The expression (5.25) was evaluated at 64 equally spaced mesh points on $[0,2\pi]$, at T=1, to generate the terminal data. The backward beam method was then used with $\Delta t = \frac{1}{301}$, and with Fourier techniques to discretize the space variable, using 64 equally spaced points on the period interval $[0,2\pi]$. The computations were performed in <u>single precision</u> on UNIVAC equipment at the University of Wisconsin. Thus, the unit round-off error is of the order of 10^{-8}. Using M= .1, we then obtain $k = \log(\frac{M}{\delta}) \approx 18.7$.

Since the exact solution is known, a comparison of the computed solution with the exact solution was made after each successive iteration. At each time $t = n\Delta t$, $n=1,2,\ldots,N$, the relative error in the discrete spatial L^2 norm, was computed after each iteration, to observe the behaviour of the backwards iteration. This error is tabulated in Table 1, for the first six iterations, for 17 values of t lying between zero and 1. Little change in the relative error pattern appears after the fourth iteration. Despite the moderately large Reynolds number of 126 in the present example, and the appearance of the exponential factor exp[Re] in the Hölder estimate for Burgers' equation, we see that even with δ of the order of $.1\times10^{-8}=10^{-9}$,

TABLE 1

Relative error in the L^2 norm, as a function of time and number of iterations, in the backwards computation of Example 1.

No of items / TIME	1	2	3	4	5	6
.0332	.29+00	.29+00	.29+00	.29+00	.29+00	.29+00
.0664	.95-01	.89-01	.85-01	.84-01	.84-01	.84-01
.0997	.54-01	.43-01	.35-01	.30-01	.30-01	.30-01
.1661	.46-01	.34-01	.26-01	.21-01	.21-01	.22-01
.2326	.42-01	.28-01	.22-01	.19-01	.19-01	.19-01
.2990	.37-01	.22-01	.17-01	.15-01	.16-01	.16-01
.3654	.34-01	.21-01	.17-01	.16-01	.17-01	.17-01
.4319	.29-01	.16-01	.14-01	.13-01	.13-01	.13-01
.4983	.27-01	.11-01	.84-02	.80-02	.81-02	.81-02
.5648	.23-01	.93-02	.79-02	.77-02	.78-02	.78-02
.6312	.18-01	.75-02	.66-02	.65-02	.65-02	.65-02
.6977	.17-01	.13-01	.12-01	.12-01	.12-01	.12-01
.7641	.12-01	.29-02	.26-02	.26-02	.26-02	.26-02
.8306	.82-02	.40-02	.39-02	.39-02	.39-02	.39-02
.8970	.83-02	.75-02	.75-02	.75-02	.75-02	.75-02
.9634	.45-02	.47-02	.47-02	.47-02	.47-02	.47-02
.9967	.32-02	.32-02	.32-02	.32-02	.32-02	.32-02

a relative error of less than 10% is achieved as far as 93% of the way back from T=1. In the above example, the exact solution does not develop steep gradients within the time interval [0,1]. Moreover, the forward iteration converges, although (5.11) is violated.

Example 2

The problem is again (5.22)-(5.24) but with $u_0 = 40$ and $\nu = 1$. As in the previous example,

$$Re = 126 \tag{5.28}$$

However, in lieu of (5.11), we now have

$$\left(\frac{64LT}{\nu}\right)^{\frac{1}{2}} \| A^{\frac{1}{2}} a \| \approx 711. \tag{5.29}$$

In this example, the forward iteration diverges for $t \geq .04$. An independent evaluation of the exact solution, using (5.25), reveals that the initial sine wave develops steep gradients almost immediately. See Figure 1. A smooth approximation to a "shock" evolves, broadens, and then dies out. Attempts were made to compute this problem backwards in time, in single precision, starting from a time T_1 sufficiently close to zero, that the "shock" would still be clearly defined in the terminal data. This amounts to reconstructing steep gradients after they have been smoothed, and clearly requires considerable precision in measurement. No measurable success was achieved in this experiment.

Although Example 1 has the same Reynolds number as the present example, it appears that the exponential factor in the stability estimate plays a much more important role in the present problem.

Example 3

We now consider an inhomogeneous problem,

$$u_t = \nu u_{xx} - uu_x + 9\pi e^{-8\pi^2 \nu t} \text{Sin } 4\pi x, \quad 0 < x < 1 \tag{5.30}$$

$$0 < t < 1,$$

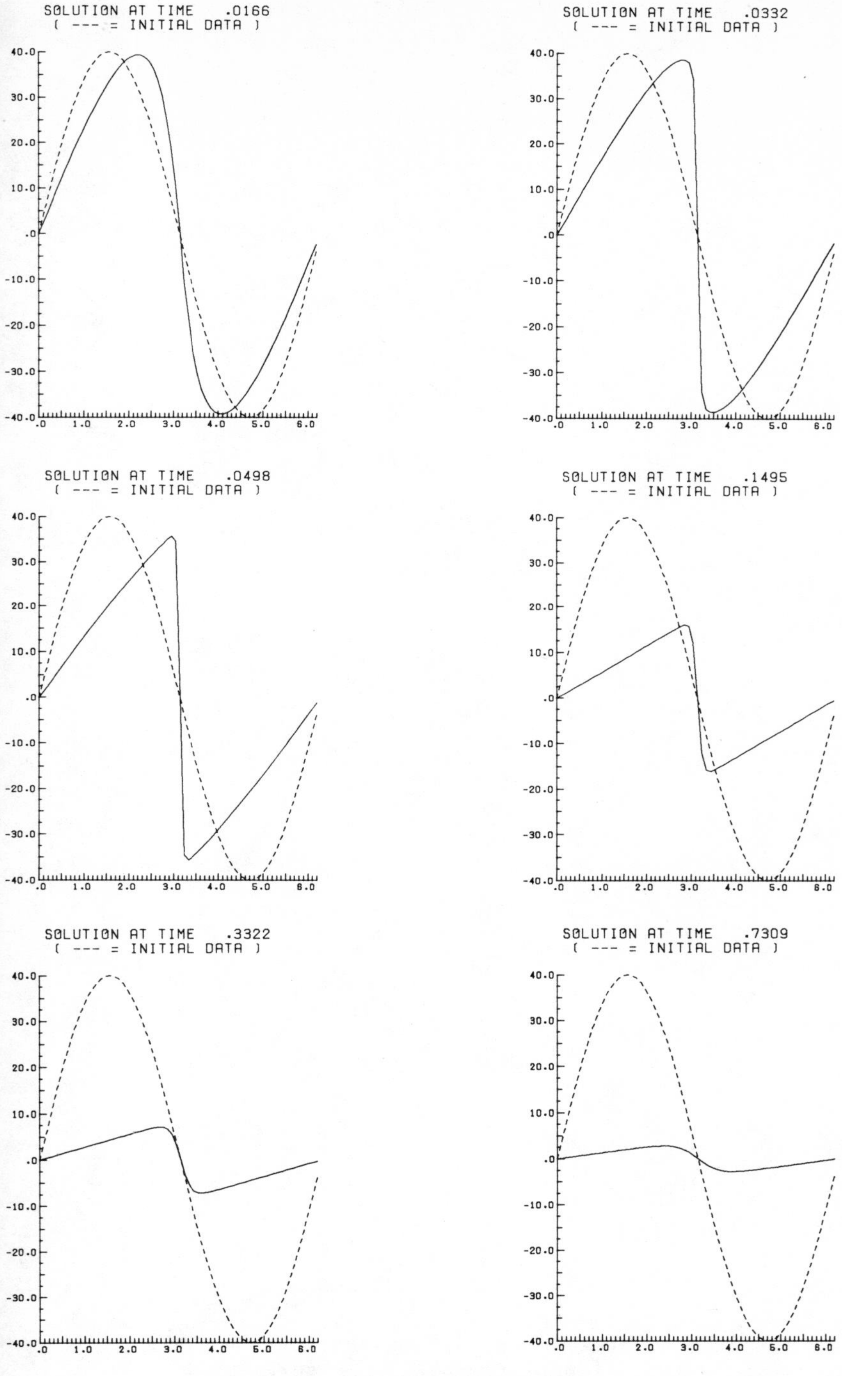

Figure 1. Development of steep gradients in the exact solution of the problem in Example 2.

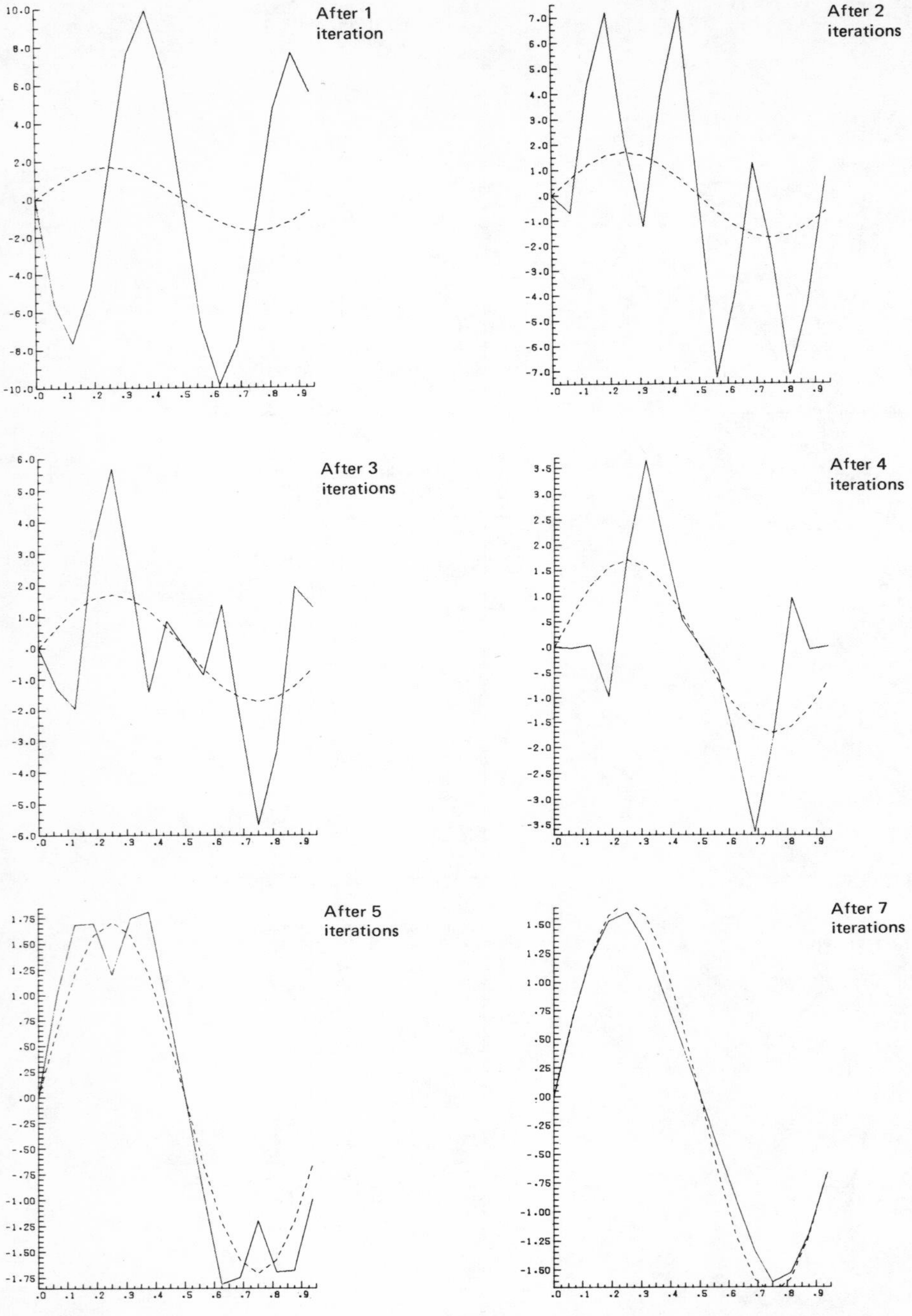

Figure 2. Convergence of the iteration at 93% of the way back from T = 1, in the computation of Example 3 backwards in time: —— = computed solution at time 0.066, (- - - = exact solution), after 1,2,3,4,5,7 iterations.

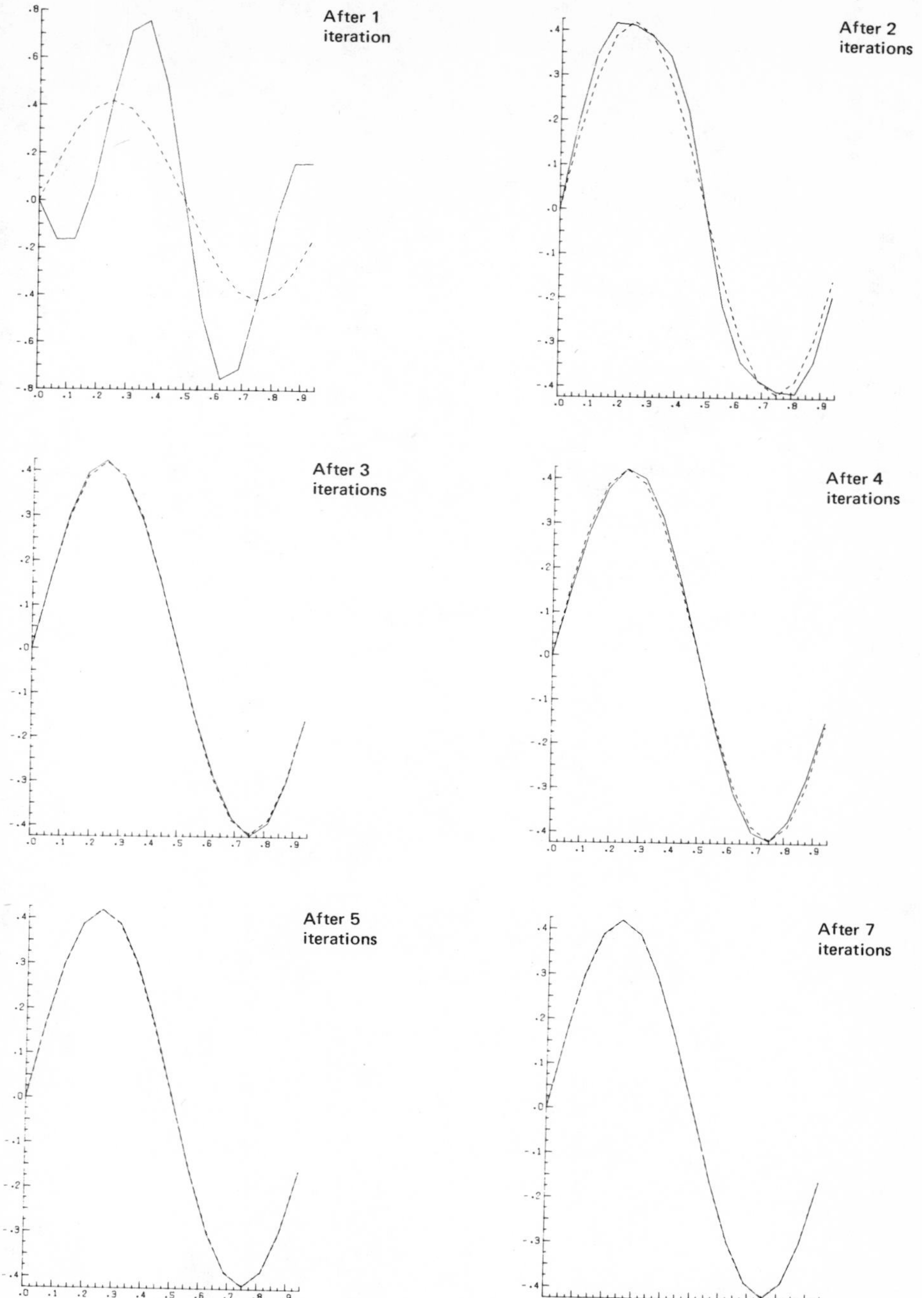

Figure 3. Convergence of the iteration at 77% of the way back from T = 1, in the computation of Example 3 backwards in time: —= computed solution at time 0.233, (--- = exact solution), after 1,2,3,4,5,7 iterations.

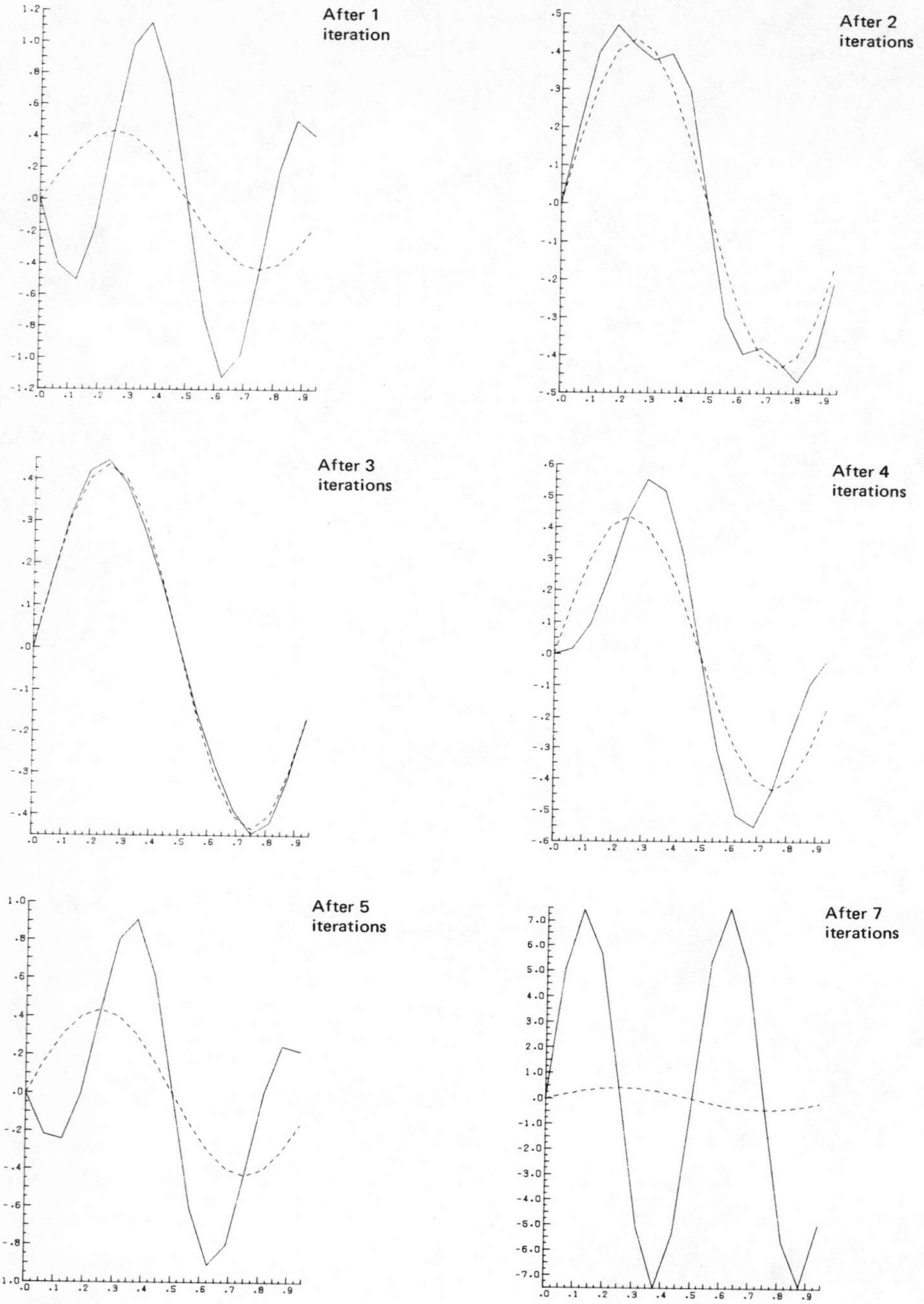

Figure 4. Asymptotic convergence phenomenon at 77% of the way back from T = 1, in the computation of Example 4 backwards in time: — = computed solution at time 0.233, (--- = exact solution), after 1,2,3,4,5,7 iterations.

$$u(0,t) = u(1,t)=0, \quad t > 0, \tag{5.31}$$

$$u(x,0) = 3 \operatorname{Sin} 2\pi x, \qquad 0 \leq x \leq 1. \tag{5.32}$$

The exact solution is

$$u(x,t) = e^{-4\pi^2 \nu t} \operatorname{Sin} 2\pi x. \tag{5.33}$$

We chose $\nu = 3/14$ so that Re = 14 in this experiment. Note however that in lieu of (5.11), we have

$$\left(\frac{64LT}{\nu}\right)^{\frac{1}{2}} \left[\| A^{\frac{1}{2}} a \| + \int_0^T \| A^{\frac{1}{2}} f(s) \| \, ds \right] \approx 487. \tag{5.34}$$

Convergence of the forward iteration occurs at this and even higher values of (5.34) in this type of example. However, even here, where the solution does not develop steep gradients, divergence of the <u>forward</u> iteration occurs when $\nu = .05$, in which case (5.34) has a value of 2700. See [6]. The importance of the value of (5.34), rather than the Reynolds number, is quite apparent in the behaviour of either the forward of backwards iterations, and this remains valid in all our experiments.

In the backwards computation of this problem, the relative error at 93% of the way back from T = 1, was found to be of the order of 500% after the first iteration! This error was then reduced to less than 10% after six iterations. At 77% of the way back from T = 1, the initial relative error of 110% was reduced to less than .3% after six iterations. In Figure 2, the first seven iterates are plotted, together with the exact solution, at t = .0664, the 93% value. In Figure 3, the iterates at 77% of the way back are depicted. Again the influence of the exponential factor in the convexity estimate, does not seem to be present in this single precision computation. Indeed, the accuracy which one can achieve at such long distances into the past is very encouraging.

4. AN EXAMPLE OF "ASYMPTOTIC" CONVERGENCE

In Example 3, the value of 487 in (5.34) is somewhat of a critical value insofar as observing the divergence phenomenon suggested by Theorem 5.2, in less than eight or nine interations. We consider now the problem in Example 3 with a slightly lower value of ν, $\nu=3/14.256789$. We then have,

$$\left(\frac{64LT}{\nu}\right)^{\frac{1}{2}} \left[\|A^{\frac{1}{2}}a\| + \int_0^T \|A^{\frac{1}{2}}f(s)\| ds\right] = 496. \qquad (5.35)$$

In Figure 4, the first seven iterates at 77% of the way back from T = 1 are depicted. The third iteration gives the closest agreement. A more detailed discussion of this and other features of the algorithm is given in [6]. The forward iteration converges in this example.

REFERENCES

1. S. Agmon and L. Nirenberg, Properties of solution of ordinary differential equations in Banach space, C.P.A.M. 16 (1963), 121-139.
2. B.L. Buzbee and A. Carasso, On the numerical computation of parabolic problems for preceding times, Math. Comp. 27 (1973), 237-266.
3. A. Carasso, The abstract backward beam equation, SIAM, J. Math. Anal. 2 (1971), 193-212.
4. A. Carasso, The backward beam equation: Two A -stable schemes for parabolic problems, SIAM J. Numer.Anal. 9 (1972), 406-434.
5. A. Carasso, Error bounds in the final value problem for the heat equation, MRC Technical Summary Report // 1479, September 1974, Mathematics Research Center, University of Wisconsin-Madison, Wis. 53706.
6. A. Carasso, Computing small solutions of Burgers' equation backwards in time. MRC Technical Summary Report // 1525, January 1975, Mathematics Research Center, University of Wisconsin-Madison, Madison, Wis. 53706.
7. J.D. Cole, On a quasilinear parabolic equation occurring in aerodynamics, Quart. Appl. Math. 3(1951), 225-236.
8. R.E. Ewing, The approximation of certain parabolic equations backwards in time by Sobolev equations, SIAM J Math. Anal. (To appear).

9. F. John, Numerical solution of problems which are not well-posed in the sense of Hadamard, Proc. Rome Symp. Prov. Int. Comp. Center (1959), 103-116.

10. T. Kato and H. Fujita, On the non-stationary Navier-Stokes system, Rend. Sem. Mat. Univ. Padova 32 (1962), 2 243-260.

11. T. Kato and H. Fujita, On the Navier-Stokes initial-value problem, Arch. Rat. Mech. Anal. 16 (1965), 269-315.

12. R.J. Knops and L.E. Payne, On the stability of solutions of the Navier-Stokes equations backwards in time, Arch. Rat. Mech. Anal., 29 (1968), 331-335.

13. H.O. Kreiss and J. Oliger, Comparison of accurate methods for the integration of hyperbolic equations, Technical Report, No. 36, Department of Computer Sciences, Uppsala University, October 1971, Uppsala, Sweden.

14. K. Miller, Nonunique continuation for uniformly parabolic and elliptic equations in self-adjoint divergence form with Holder continuous coefficients, Arch. Rat. Mech. Anal. 54 (1974), 105-117.

15. S.A. Orszag, Numerical simulation of incompressible flows within simple boundaries I, Studies in Applied Mathematics 50 (1971), 293-327.

16. L.E. Payne, Improperly posed problems in partial differential equations, Lecture Notes, National Science, Foundation Regional Conference on Ill-Posed Problems, May 1974; Department of Mathematics and Statistics, University of New Mexico, Albuquerque, N.M 87131.

17. R.D. Richtmyer and K.W. Morton, Difference Methods for Initial Value Problems, 2nd Ed., Interscience, New York (1967).

18. J.N. Franklin, On Tichonov's Method for Ill-Posed Problems, Math. Comp. 28 (1974), 889-907.

Sponsored by the United States Army under Contract No. DA-31-124-ARO-D-462.

University of New Mexico
Albuquerque, New Mexico
USA

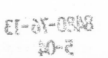